Wolves and People

The Management Imperative and
Mythology of Animal Rights

Jim Hatter

by

James Hatter, Ph.D.

© Copyright 2005 James Hatter.
All rights reserved. No part of this publication may be reproduced, stored in a retrieval system, or transmitted, in any form or by any means, electronic, mechanical, photocopying, recording, or otherwise, without the written prior permission of the author.

Also by James Hatter:
Politically Incorrect

Note for Librarians: A cataloguing record for this book is available from Library and Archives Canada at www.collectionscanada.ca/amicus/index-e.html
ISBN 1-4120-6147-4

Printed in Victoria, BC, Canada. Printed on paper with minimum 30% recycled fibre. Trafford's print shop runs on "green energy" from solar, wind and other environmentally-friendly power sources.

TRAFFORD
PUBLISHING

Offices in Canada, USA, Ireland and UK

This book was published *on-demand* in cooperation with Trafford Publishing. On-demand publishing is a unique process and service of making a book available for retail sale to the public taking advantage of on-demand manufacturing and Internet marketing. On-demand publishing includes promotions, retail sales, manufacturing, order fulfilment, accounting and collecting royalties on behalf of the author.

Book sales for North America and international:
Trafford Publishing, 6E–2333 Government St.,
Victoria, BC v8t 4p4 CANADA
phone 250 383 6864 (toll-free 1 888 232 4444)
fax 250 383 6804; email to orders@trafford.com

Book sales in Europe:
Trafford Publishing (UK) Ltd., Enterprise House, Wistaston Road Business Centre, Wistaston Road, Crewe, Cheshire cw2 7rp UNITED KINGDOM
phone 01270 251 396 (local rate 0845 230 9601)
facsimile 01270 254 983; orders.uk@trafford.com

Order online at:
trafford.com/05-1048

10 9 8 7 6 5 4 3

About the Author

Born in Victoria, British Columbia in 1921, Dr. James Hatter received his early education at Lake Cowichan where he graduated from high school in 1939. He attended Victoria College and went on to graduate from the University of British Columbia in 1945. In 1947 he became British Columbia's first game biologist. He attended Washington State University and graduated with a Ph.D. in 1952 after studying the ecology of moose in central British Columbia.

In 1952 Hatter became the province's Chief Game Biologist; and then its Fish and Game Branch Director in 1963. He also served as part-time Assistant Professor of Zoology at UBC from 1952 to 1955.

He is the author of many articles and papers on game administration and management. His interests have focused on educating the public about the factors that control the presence and abundance of game animals and other wildlife.

Jim Hatter has been a serious hunter and angler as well as a naturalist most of his life. He retired in 1979 after 32 years of public service. Jim has travelled widely; and continues to be involved in wildlife-related activities.

Acknowledgements

I am especially thankful to my son Ian Wesley Hatter who reproduced my rough hand written pages into an acceptable manuscript.

My son Dave Hatter read and studied the manuscript for legal considerations for which I am most thankful.

Georgiana Ball, editor of my book *Politically Incorrect*, offered advice and helpful suggestions for which I am grateful.

To my wife Olive I owe the most and especial thanks for her patience in allowing me the uninterrupted quiet hours to think, read and write *Wolves and People*.

Special Recognition

The author recognizes three people for their understanding and attention to the management imperative in respect to predator-prey systems:
The late Cyril Shelford, B.C. Minister of Agriculture
Tony Brummet, B.C. Minister of Environment
Douglas Janz, Former Regional Wildlife Biologist for Vancouver Island

Table of Contents

1. Preface 1
2. Introduction 3
3. Early Wolf Control – Failure of the Bounty System 5
4. Excerpts from Wolf Predation in the North Country: 1930 – 1948 8
5. Effective 1080 Wolf Control: 1950 – 1965 12
6. Government Refusal to Re-instate the 1080 Program in 1978 20
7. The Muskwa and Kechika Wolf Management Research Project – 1983 33
8. A Report from the Guide Outfitters in British Columbia – Wolf Predation in the 1990's 48
9. The Wolf Management Imperative 66
10. Wolf Predation on British Columbia's Red-Listed Mountain Caribou 71
11. Evolution, Genetics, Religion and Culture 73
12. Nature Worship – A Scientific Religion 79
13. Mythology of Animal Rights 82
14. The Wolf's Role in Evolution, Genetically Induced Phobias and Cultural Dilemmas . . . 91
15. Understanding the Human Predator: Why We Hunt, Fish and Trap 94

16. Two Accounts of Individuals Treed by
    Wolf Packs in Northern B.C. 97
17. Conclusions and Comments. 101
18. References. 103
Appendix 1. Wolf Questionnaire to Guide-Outfitters in B.C.. . . 105
Appendix 2. Letter to Pesticides Directorate, Ottawa. 110
Appendix 3. Response from Pesticides Directorate. 111
Appendix 4. Rio – The Pet Alaskan Wolf. 112

1
Preface

Wolves and People is about conservation or management of valuable ungulate species for economic and social benefits while maintaining tolerable levels of wolf predation to achieve this purpose.

Wolves and People describes the serious obstacles incurred from animal rights activists (antis) in urban society who consider the wolf an icon, unlike other species, and therefore apart from the overall objectives of wildlife conservation. However, the definition of antis who oppose hunting, fishing and trapping, does not include environmentalists (enviros) who understand and help to promote the importance of conserving and enhancing the natural environment for economic and social reasons. The majority of hunters are environmentalists who support and promote these endeavours.

Wolves and People also discusses the propensity or inclination we have to knowingly "believe in things that are not so." Many people are attracted to unscientific mythical concepts. Much of this is the result of our human failure to understand the inherent implications of our early evolutionary origins which extend back for millions of years. It is not properly understood that we carry the genes of Palaeolithic ancestors. Although we physically resemble our primitive relatives, we have been "bent out of character" by culture. Culturally acquired attitudes and beliefs are not the result of genetic evolution and inheritance. It is not culture that attracts us to the natural environment wherever we live, but rather our stone age, unshakeable inheritance from being hunters and gatherers when only nature was everything and everywhere. I argue that to believe otherwise is mythical, unnatural, unscientific and not in keeping with intellectual society.

My purpose therefore, in Wolves and People is to explain that hunting, fishing and trapping are genetically entrenched in our evolution and a natural part of our outdoor heritage. We must not allow animal rights mythology to prevail in modern society.

This is an appeal to all who hunt, fish and trap, as well as non-participants, to reject the mythical beliefs that blindly oppose all predator control, hunting, fishing and trapping.

2

Introduction

Wolves and People supports wolf control as an integral part of sustainable ungulate conservation. It bears no similarity to *Never Cry Wolf* by Farley Mowat (1963) which fictionalized the wolf and muddied the waters of conservation thinking for decades. *Never Cry Wolf* was reviewed by two professional biologists, one of whom compared it to the fable *Little Red Riding Hood*. Nevertheless, it became a best seller with over a million copies sold and won international acclaim despite its depiction of fiction as fact. In British Columbia, it was placed in all high schools in the mid sixties as optional reading for grade 9 students. Some teachers believed it to be a true account of the experiences of the author and made it compulsory reading in grade 9 English.

Wolves and People explains why wolf control has become a controversial issue especially among urban anti-hunting and animal rights proponents. History has demonstrated that without science-based control of wolf predation and concern for the people affected by it, they may take matters into their own hands, resulting in unnecessary inhumane treatment of wolves and other predators.

While I have respect for authors of the many books about wolves I am critical of those whose prejudices have knowingly elevated the wolf in the urban mind to the status of an icon. This has happened without thought to the fact that mankind is the dominant predator in nature's food chain. The human predator has an inherited right to control competing predators. I argue that to believe otherwise is the result of cultural corruption, ignorance of genetic inheritance from ancient ancestors and the human failure to "believe in things that are not so." Governments often respond to emotional urban minorities who use the press to oppose wolf population control, believing that it is socially unacceptable. In British Columbia, 80 percent of voters live in the urban or city environment. This large segment of society is vulnerable to the fallacious propaganda of animal rights advocates who oppose all hunting and predator management.

It is evident that controversial issues sell more newspapers when there is conflicting and polarized public interest in wildlife conservation. Informative contacts between professional wildlife biologists and the press have been discouraged or even disallowed by recent governments in British Columbia. As a result, opposition to wolf control has become an issue of political concern.

Because evolution and culture have left us with the propensity to unscientifically "believe in things that are not so" the writer has ventured into related aspects of nature, culture, mythology and religion.

3

Early Wolf Control – Failure of the Bounty System

In the early years of the fur trade and the Hudson's Bay Company in British Columbia, there were some attempts and incentives to control wolves to protect domestic livestock of the early settlers.

In the United States when only bounties were in effect the wolf is said to have held back development of the sheep industry in the state of Virginia for most of the seventeenth century, as well as seriously affecting expansion of the cattle industry westward across the continental grasslands. It was believed that the eastern United States would never be an important sheep country because of the abundance of wolves. Extirpation appeared to be impossible even though bounties were paid continuously for more than two centuries. The State of Virginia paid wolf bounties from 1632 to 1939 – a period of 307 years. In the final analysis, it was not bounties that brought about the elimination of wolves but rather the destruction of their habitat by the farming industry and the widespread use of strychnine poison (Young, S.P. and E.R. Goldman. 1944. *The wolves of North America*).

In 1905, A. Bryan Williams, appointed the first Provincial Game and Forest Warden in British Columbia soon became aware that wolf predation had become a political problem. Accordingly, the bounty system was endorsed by the government of the day.

For the next forty-five years the only involvement of government in wolf control was through payment of bounties. Unlike the American States, the use of poison was strictly prohibited. It was a serious offence under the Game Act to even posses poison.

The control of wolf predation in British Columbia under the bounty system was not generally effective. Although it was against the law to use strychnine to kill wolves to collect a bounty, it was done to some extent and bounties were paid on wolves that had been poisoned illegally. To collect the bounty, it was necessary to cape out the wolf's head, or bring in the entire pelt for examination by the clerk on duty at the nearest Government Agents office. In the summer months the capes or hides were often subject to decay, covered by maggots and in a state of putrefaction. There are stories about office staff refusing to examine putrid capes and hastily paying off a bounty applicant with his bag of skins from a litter of domestic pups alleged to be young taken from a wolf den or more commonly from coyote pups.

The incentive to hunt or trap wolves resulted in bounty hunters going to areas in which wolves were most abundant and easier to hunt rather than areas where problem wolves were menacing domestic stock. In 1947, bounties were paid on 1,102 wolves. An increase from $10 to $25 in the Cariboo district did not result in any more wolves being killed. In 1948, in the Cariboo, Lillooet and Kamloops districts, the bounty was further increased to $40 due to complaints from cattleman.

Throughout the period from 1905 to 1950 some people did obtain access to poison and used it illegally to prevent predation on domestic livestock and big game animals.

The late George Ball, a pioneer guide and outfitter at Telegraph Creek in northern B.C. in the 1930's and 40's was alarmed at the extent of wolf predation on Stone's Sheep and other trophy big game animals in his guiding territory. An American, who was a doctor, mailed him small amounts of strychnine poison which he discreetly distributed to other local guides who he could trust. His daughter, Georgiana, remembers the occasion when their home and out-buildings were searched for illegal possession of poison. The word was out that her father was using strychnine. He received a hasty warning that a British Columbia policeman was on his way to search for illegal possession of poison. George had only a short time to relocate his small bottles of strychnine from the usual hiding place where a careful search would discover them.

It was mid-winter, with a fair depth of snow at the home ranch. In the horse pasture not far from the house, 50 head of horses and persons feeding them had trampled the snow wherever hay was spread out. Into a small pile of snow George Ball hid his small bottles of powdered strychnine. His buildings were carefully searched but the ingenious hiding place in the horse pasture was not suspect.

4

Excerpts from Wolf Predation in the North Country:

1930 – 1948 by L.P. Callison

In 1948, an American hunter named L.P. Callison of Seattle, Washington published a booklet of 89 pages entitled Wolf Predation in the North Country. He travelled extensively in Alaska, Yukon and northern B.C. from 1938 to 1948 and before that to a lesser extent in the late twenties and early thirties. He also sent questionnaires to guides and others to gather information on the increasing amount of wolf predation in the north in 1948 compared to earlier years. Following are some remarks he received.

C.A. Sheffield and Henry G. Courvoisier, trappers in the Muskwa

"In comparing today (1948) with the game of 1930 in this territory, I'd say there was a vast decrease in sheep, caribou and moose. In 1930 game parties could see bands of sheep from fifty to one hundred; today they see from ten to fifteen. In the lowlands, where trails were two feet deep, today are all grown up with moss. We see very few calves or yearlings or lambs and the wolf stools along the trails are full of hair and bone. This covers the full drainage of the Muskwa River to where it empties into the Nelson River at Fort Nelson. The kills I have found by wolves in my twenty-two years of trapping are too numerous to count. Moose, I say, about a quarter today what there were ten years ago in 1938. Sheep about half. Now, since the moose are mostly killed off, the wolves are moving to the mountains and slaughtering the caribou and sheep. There are very few deer in this area now. Moose, I would say, about a quarter today than there were ten years ago. Sheep about half."

J.E. Clark, North Tweedsmuir Park

"Wolves definitely on the increase. Fifteen or even ten years ago it was a rare sight to find even a lone wolf track and now, it is a common occurrence to find the tracks of packs of six or eight to sixteen to twenty wolves. I have seen over one hundred fifty head of moose this winter that I have reason to believe were different animals and did not see one moose calf, whereas a few years ago calves would have been around twenty percent."

James R. Stanton, Knights Inlet

"Wolves on the increase. Timber wolves are taking an estimated two thirds of the increase in Grizzly bears, which they accomplish by snatching the first year cubs."

Rene G. Dhenin, Peace River

"Wolves very much on the increase and if something is not done about it they will kill all our game off or drive it away."

Dick Church, the Chilcotin

"The wolves are numerous in this district and their ravages are already being felt on the game. Wolves are on the increase."

Wolves and People

Thomas B. Armstrong, West and North of Quesnel

"Twenty years ago there were few wolves. They started coming in about ten years ago."

Clifford B. Eagle, Lac la Hache

"Twenty years ago lots of caribou and very few wolves. Today, no caribou and plenty of wolves."

Ted Helset. Clearwater, South Wells Gray Park

"Fifteen years ago caribou were plentiful. It was not unusual to see herds of from twenty to forty. At that time one seldom saw sign of wolves. Today the caribou are almost extinct whereas the wolves have increased to such an extent that they are running in packs of as many as twenty."

James P. Cochran, Barkerville

I moved into the unsettled area north-east of Barkerville in 1912, and have been guiding, trapping, prospecting and picture taking ever since, being in the game country both summer and winter. The wolves started to come in bunches in 1932. Since then, they have destroyed, I would estimate, eighty-five percent of the caribou, thirty-five to fifty percent of the moose, twenty percent of the deer and goats."

H. M. McNeill, Central Cariboo, Fawn Post Office

"There are probably five wolves now to one four years ago. There used to be many caribou on the high mountains – it was big caribou range. At that time there were no wolves. Now there are many wolves in that area in the summer when I have been there. But the caribou are gone. It seems reasonable to believe that the wolves have devoured them."

L. P. Callison summed up the effect of wolves on caribou by saying; "When measured by the number of victims it is the caribou that have suffered most in the North Country. Their vast numbers, their dumb curiosity, and their lack of any effective offensive or defensive armour make them easy prey for the wolves. Conservative estimates

The Management Imperative and Mythology of Animal Rights

place the loss during recent years in the millions." His concluding remark was, "You can feed the wolves or you can feed the human predators but you can't feed both on wild game."

5

Effective 1080 Wolf Control: 1950 – 1965

Not until 1950 was it possible to control wolf and coyote predation in British Columbia. At the first Annual Game Convention held at Harrison Hot Springs Hotel in 1946, irate Chilcotin ranchers came with the request for more effective control of wolves attacking their cattle. Again at the 1947 Convention, the demands were repeated, this time in the presence of Attorney General Gordon Wismer and the Provincial Minister of Agriculture. It was apparent that the Game Commission favoured more wolf control but to allow the use of poison, as in the United States, was not something Commissioners Butler and Cunningham wanted to pursue without government approval.

There are four reasons why intensive wolf poisoning began in the 1950's. Firstly, the cattlemen carried considerable influence on government. Secondly, the Game Commissioners were aware that the adjacent State of Washington had a poison program for coyotes that was accepted publicly. Thirdly, a comparatively humane canid poison, sodium monofluroacetate or 1080 had been developed by Monsanto Chemical Company to be less lethal to non-target species of wildlife. And finally, its use would be restricted to trained staff under direction of a Chief of Predator Control.

By 1950, British Columbia had a separate Predator Control Division, responsible to the Game Commission, similar to that of Washington State. W. Winston (Bill) Mair was appointed Chief of Predator Control in charge of 12 predator control officers.

Predator Control Officer's Uniform Insignia

In 1951 a rabies outbreak occurred in northern Alberta, causing fears that this serious disease would spread to British Columbia. The Federal Department of Agriculture ordered destruction of wolves and coyotes west of the Rocky Mountain passes, through which rabid animals could enter British Columbia. The Finlay, Gataga and Kechika river valleys, known to be heavily populated with wolves, were intensively poisoned from the air with 1080 baits in 1952, 1953 and even more heavily in 1954.

This task fell to the late Milton (Milt) Warren, a decorated World War II veteran, who in 1950 was hired as a game warden to become the main participant in the ensuing wolf control program in northern British Columbia. Following are the details of the control program, known to the writer. Nowhere in the records are there details of Warren's 1080 program to initially prevent the spread of rabies into BC. In several interviews with him in 2000, he told me that the lack of detailed records or published information was not for political reasons but rather a failure of his supervisors to work up reports in any detail.

Wolves and People

Bill Mair and later Al West, his supervisors, never asked for written reports or observations from which to prepare permanent government records. He was required only to submit flight maps with the location of bait stations marked with an 'x'. "Really, after that there was little I could do without money for follow-up observations" Warren said.

I believe Warren's work was as well-executed as could be expected under the prevailing circumstances. From my several detailed interviews with him, it was obvious that he possessed good judgment and common sense. He told me that the objective, in addition to blocking the westward spread of rabies was to bring about a better wolf-prey balance to benefit both ungulates and the hunter. Although the poison program in the north was 75 percent to enhance ungulate management there were also important benefits to be gained from helping ranchers located in wolf inhabited areas.

Baiting was done only once a year on high density winter ranges – "a one shot baiting for the year," recalled Warren, "with fewer baits after 1954." Most were dropped on snow covered frozen lakes and rivers from an altitude of about 200 feet. Wolves would dig them out from under the snow or when covered by overflow ice. Baits were also dropped on clear ice when thick enough to hold them.

Warren trained and instructed other predator control officers in the use of 1080. Most aerial baits were horse meat although road killed game animals were also used to a lesser extent. One year, 75 horses were processed into wolf baits. One horse carcass provided twenty-five 30 lb. baits. This size was necessary to last until break-up. Fifteen to twenty baits could be carried in a supercub, used extensively because of the low cost. Warren had very limited operating funds and it was necessary to make them last to the end of each winter baiting season.

The elimination of entire wolf packs was not uncommon. Warren estimated that it took only 2 oz. of treated meat to kill a wolf compared to over a pound for a person. He prepared his own 1080 baits, using a teaspoon of 1080 powder to one gallon of warm water. This solution was injected into quarters of a freshly killed horse, with a large syringe called a brine gun used to impregnate pork with a salt solution in the preparation of ham. Each quarter was treated several times to insure a good dispersion of solution throughout the flesh.

Unfortunately, 1080 is a slow acting poison which may enable a wolf or coyote to travel well away from a bait station before dying. This makes it difficult to record the number killed. Poison 1080, however, is highly preferable to strychnine as evidenced by the fact that when the victims are found there is little or no evidence of a struggle.

Strychnine and to a lesser extent potassium cyanide are not considered to be humane poisons. They were widely used in control of wolves and coyotes before the invention of 1080. Some strychnine and potassium cyanide was used by Warren but 1080 was the predominant choice. His method of preparing strychnine baits followed a simple procedure. He used a heavy piece of meat or a carcass to prevent the predator from carrying it off, thus assisting in later pickup of the unused portion. The large bait was lacerated with deep cuts into which was rubbed small amounts of coarse crystalline strychnine.

George Ball and his trapper friends in the Cassiar placed their strychnine treated moose or caribou meat baits, covered in hide, into water filled holes cut through lake ice at wolf crossing points. They were left to disappear at break-up. Only wolves and ravens were killed.

Unlike 1080, strychnine has a taste, making it less attractive to a wolf. An amount equivalent to the volume of an aspirin was considered adequate to kill a wolf. Unlike 1080, the victim dies close to the bait station. It is not considered a humane poison as there may be violent throes before the animal dies. Warren has found as many as 10 or 12 wolves within 100 metres of a strychnine bait. In 1947, the author visited a strychnine bait station on 9 Mile Lake near Quesnel. Around it were seven dead wolves, not more than 50 metres from the bait.

Potassium cyanide capsules dropped onto lakes from aircraft were also used in wolf control in the 50's, sometimes in conjunction with 1080. Double ought (00) gelatine capsules were employed to contain the poison. The loaded capsules were then smothered with moose tallow and bone marrow and covered with thin paper until used. This helped to conceal the odour of the chemical.

Alfred Harrison, a resident at Ootsa Lake is believed to have been the first person to use potassium cyanide pills to kill wolves, according to Warren. Cyanide pills, strychnine and 1080 baits were dropped on frozen rivers and lakes only, to avoid poisoning non-target animals, such as small furbearers.

Warren found that wolves become suspicious and avoid strychnine and potassium cyanide baits because of odour but not so with 1080. It has no taste or odour. Warren recalled placing 10 cyanide stations on the Torpy River. When he returned to check the pill locations, he found that wolves had defecated on all of them. He discovered that alpha male and female wolves learn to avoid strychnine and cyanide baits but young wolves are susceptible. Skunk scent or fish oil was used to attract wolves to the small, somewhat invisible pills containing potassium cyanide.

Warren told me about the downside to the aerial poison program in the fifties. Eagles that returned early and fed on his 1080 baits along with the ravens were vulnerable to poisoning. It was therefore necessary to pick up all unused land baits on the last inspection. He agreed that strychnine and 1080 will repeat if sufficient stomach contents are vomited up and eaten by a susceptible animal.

It is not surprising that Milt Warren experienced several accidents during his years as a predator control officer. In responding to individual complaints to protect livestock in the Burns Lake area, several dogs were poisoned. These included those of the local doctor and policeman. Milt said he "collected" replacement dogs on this occasion.

Another time several dogs died on landing strips from melting of frozen 1080 baits in the aircraft. Little wonder there were accidental mishaps when Warren was forced to operate each year on a meagre amount of $2000 from regional budgets for flying expenses.

In 1952 and 1953, the 1080 baits had to be stored in a wooden shed in Prince George due to spending constraints for safer facilities. Incredible as it may seem, someone hired a taxi and stole baits from the shed during daylight hours. They were then hidden in a snow bank with the result that local dogs began dying. Three men actually ate the poisoned horsemeat but suffered only violent diarrhoea attacks. The 1080 concentration strong enough to kill wolves was not potent enough to kill a person. On another occasion, a nine year old girl ate some poisoned meat that had been cut from a huge bait. Warren was

The Management Imperative and Mythology of Animal Rights

worried that her small size might prove fatal but she survived the incident. Her discomfort was apparently unrecorded.

On the Osilinka River 5 or 6 dogs died in an Indian dog team. On another occasion Harold Moffat, mayor of Prince George, lost his black Labrador to a poison bait. This happened when the dog picked up small scraps and blood that were eroded from the baits as they were carried from a truck into the storage shed. Milt replaced the dog! He was surprised another time when a lynx died of 1080 when it ate enough meat to poison it. The cat family is more resistant to 1080 than canid animals for which it was developed.

At Cold Fish Lake where 1080 baits were dropped, the Indian guides refused to eat the wild berries or drink the water.

Warren and wildlife biologist Ken Sumanik became proficient at estimating the size of a wolf pack from their tracks in the snow before the aircraft caught up with the animals. This knowledge was helpful in estimating potential mortality from baits dropped in an area frequented by a wolf pack. Ken Sumanik was opposed to wolf control as were other biologists of those earlier years. I do not recall Sumanik's account in the 50's describing a potential method of determining wolf mortality from tracks only when the animals were not visible at the time the bait was dropped.

It was unfortunate and of much concern to Warren and to Sumanik too I expect, that no funds were made available for follow-up flights to determine more accurately the results of the baiting program. Warren's interest and curiosity, however, motivated him to question bush pilots and aircraft owners about the number of wolves they observed before and after baiting flights. He estimated that the aerial baiting program accounted for the removal of about 600 wolves.

There is no doubt that the wolf control measures of the 50's were effective. Well before 1962 complaints of wolf predation from the areas covered in the 50's had diminished markedly. Wolf control was scaled down to site specific local areas and in the mid-sixties we terminated all use of poison to control wolves, except when it was necessary to deal with local losses to domestic livestock. The 12 member Predator Control Division was deployed to enforcement and other activities of the Fish and Wildlife Branch. At headquarters in Victoria we believed

Wolves and People

that because the wolf and coyote control program had been so successful we could always reintroduce it if the need again arose.

When the program ended, Warren's knowledge of wildlife, along with his enthusiasm, personality and ability to meet and address the public resulted in a promotion to Information and Education Officer for the north. He organized the province's Conservation Outdoor Recreation Education Program (CORE) out of Prince George, training and supervising instructors. He became highly popular as a speaker in the classroom of grades 9 and 10 students. Unlike the meagre operating budget for wolf control he received much better funding for his educational work. His personal collection of large coloured slides used for school presentations were in demand not only at schools, but also when asked to attend girl guide, boy scout and other meetings. He put in much overtime, working form a calendar with bookings that covered extended periods of time with few breaks. His vehicle and travel expenses as before were never held up. From headquarters in Victoria he was well supported. He reported directly to Rod Cameron, head of the Information and Education section rather than to Inspector Gill in Prince George.

The extensive baiting program in the foregoing account, initially to prevent the spread of rabies from Alberta, became more site specific towards the end of the 50's and even more so by the early 60's before 1080 control was finally terminated in 1965. The Fish and Wildlife Branch became more concerned about localized wolf-prey relationships and livestock protection.

Public opposition to any wolf control whatsoever peaked in the late 70's, early 80's, and again to a lesser extent in 2004 during the effort to protect the 30 remaining Vancouver Island marmots, the world's rarest mammal.

Presentation of the Distinguished Flying Cross By King George VI to Milt Warren, the survivor of three bomber crash landings during the second world war.

6

Government Refusal to Re-instate the 1080 Program in 1978

By 1978 wolf populations had rebounded. I was no longer Director of the Fish and Wildlife Branch. Doug Janz, the regional biologist for Vancouver Island, recommended the use of 1080 as wolves were creating havoc on black-tailed deer. Rafe Mair, the new minister gave approval, but interference from organized animal rights proponents forced him to rescind the control program even before it got underway. I was told that the Minister called the Director of Fish and Wildlife and the Chief of Wildlife Management to his office where he asked if they thought he had acted too quickly in rescinding the renewal of 1080 wolf control. Both persons answered simultaneously. The Director replied "yes" as I would have done if director but his Chief of Wildlife said "no." Undoubtedly, Rafe Mair was receiving a lot of criticism from the urban antis and was probably under pressure from his cabinet colleagues, not withstanding the serious loss of deer to wolf predation on Vancouver Island.

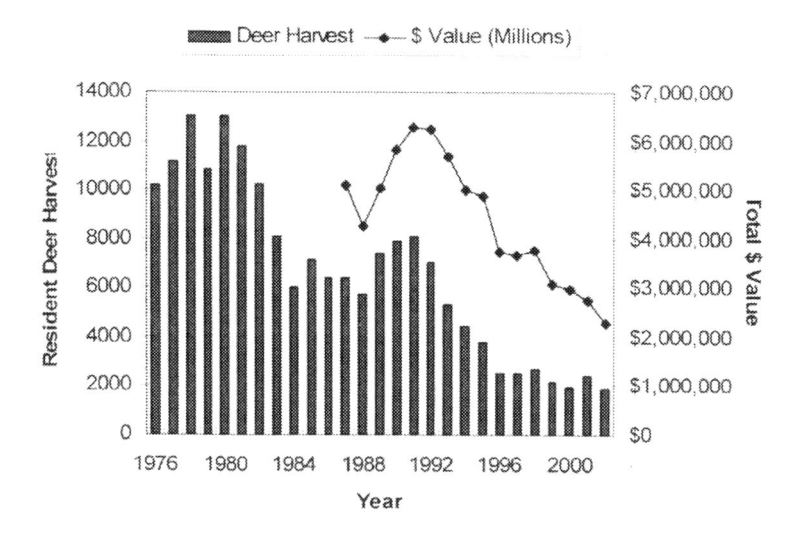

Trend in resident deer harvest and economic value on Vancouver Island.

In the above graph note how rapidly the deer harvest declined following the high wolf density in Region 1 (Vancouver Island). In eight years, the black-tailed deer harvest had plunged from nearly 14,000 deer in 1980 to only 6,000 in 1988. The increase to 8,000 animals in 1990 was a result of the less effective alternative of trapping, but it too came under opposition of animal rights proponents who were opposed to government supplying traps to kill wolves. Note the consequent decline in deer hunter expenditure from almost 7 million dollars in 1980 to less than 2.5 million in 2002.

When the Government refused to re-instate the 1080 program animal rights activists were encouraged to make it even more difficult to again use 1080. They lobbied the Government to require that a provincial, as well as the federal permit, would be required. In addition, the Fish and Wildlife Branch would be required to advertise its intention to apply for a Federal 1080 permit in the BC Gazette as well as in selected newspapers. If a federal permit were granted, approval again had to be made public in the Gazette and local newspapers. The "antis" made doubly certain they would be well advised and prepared in the event 1080 wolf control was planned. The Government unfortunately complied with their demands.

After retiring from Government in 1979, I was told that the Federal Department of Agriculture would refuse to issue 1080 permits for ungulate enhancement because it was no longer their policy to allow 1080 to be used for this purpose. I later wrote to the permit issuing office in Ottawa to verify this advice and to my surprise, I learned that a federal 1080 permit could have been issued if requested by the Province in 1978. I assume the same still applies (Please see appendix 2).

Before retiring, I was allowed to examine 377 letters received by the Government in opposition to resumption of wolf control. Only a small number of the letters were composed by school children. It was obvious, however, that some teachers had prompted students to protest a renewal of wolf control. Some letters from school children sounded like remarks the students had copied from the blackboard or read in Farley Mowat's book, *Never Cry Wolf*.

I submitted my findings in the following report but there was no response or government interest. Political correctness and ignorance prevailed.

My report, *Public Opposition to Wolf Control and the Use of* Poison (*an Examination of Expressed Concerns*) follows:

Beginning in January 1978 the office of the Minister, the Premier's office and the Fish and Wildlife Branch headquarters in Victoria received letters protesting an announcement of renewed wolf control in north-central B.C. Between January 1, 1978 and January 30, 1979 a total of 377 letters were received – six were "repeats" thus reducing the total to 371.

When were the letters written?

Only a small percentage of letters were written between January 1978 and November 1979. In December 1979, however, 46 percent of the 371 letters were produced. This clearly illustrates an organized campaign to influence government policy. It does not reflect a true expression of individual concern.

The Management Imperative and Mythology of Animal Rights

'A plaintive cry for wolves
Vanc. Express - Nov 22/78

Wolf poisoning 'utterly inhumane'
Kamloops News Sentinel Nov 27/78

'Horror death program in B.C.
Vernon Daily News Oct 24/78

Saner methods than poisoning
Colonist - Dec 5/78

Sierra Club issues warning
Int. News - Smithers Oct 11/78

Wildlife at wolves' door
Colonist - Dec 8/78

Animal Poisoning Plan 'Inhumane'
Victoria Times Feb 17/78

Poisoning wolves affects all animals
Mar. 30/78 N. Times - Terrace

Wolf poisoning cruelest of deaths
Colonist - Dec 2/78

CRUEL DEATH PLANNED FOR WOLVES
Colonist - Nov 25/78

Examples of newspaper headlines from letters opposing the proposal to reinstate the 1080 wolf control program in 1978. Please check pages 29-32 for content in some matching headlines.

23

Sierra Club lobbies against wolf poisoning

371

New Westminster Columbian - Jan 4/78

Sierra Club on side of wolves

376

Kamloops Sentinel (News) Dec 20/78

Howl of Protest Greets Wolf Program

Victoria Times Nov 24/78

Wolf moratorium likely 2 more years

371

Newspaper headlines continued 1978.

Who wrote them?

Seventy-nine percent of the letters came from British Columbia residents, 13 percent from other provinces (mainly Ontario) and 7 percent from the U.S.A. Greater Vancouver supplied 20 percent of the B.C. letters and 11 percent came from Victoria. These two urban centers contributed 31 percent of the correspondence originating from within the Province. The remainder came from about forty provincial locations, almost entirely south of Prince George. There were almost no anti-control letters from the northern half of the Province where wolves were most abundant. Sixty percent of the letters were written by women.

Organizations for and against wolf control

About thirty-five organizations within the Province were referred to by writers belonging to them. In some cases, letters were received on behalf of an Association. These associations are listed according to opposition or support for wolf control. A number of organizations did not clearly state approval or opposition but questioned the need to kill wolves to protect caribou and called for a further independent enquiry. Many naturalist clubs took this position. These organizations are listed separately.

Opposed to wolf control

1. International Ecology Society (U.S.A.)
2. Association for the Protection of Fur Bearing Animals (B.C.)
3. Canadian Wolf Defenders
4. Central Okanagan Naturalists Club (B.C.)
5. Humane Educators of B.C.
6. Society for Pollution and Environmental Control (B.C.)
7. Animal Protection Institute of America (U.S.A.)
8. Wild Canid Survival and Research Centre (U.S.A.)
9. Amalgamated Conservation Society (B.C.)
10. World Federation for Protection of Animals (Switzerland)
11. Greenpeace (B.C.)
12. Society for Prevention of Cruelty to Animals (B.C.)
13. Kelowna Jaycees

Wolves and People

14. Spruce City Wildlife Assn. (B.C.)
15. Princeton Fish and Game Association
16. Pender Island Farmers Institute
17. Sierra Club of Canada (B.C.)
18. Comox Strathcona Natural History Society
19. Victoria Natural History Society
20. Citizens Association for Predator Conservation (B.C.)
21. Ontario Wolf League
22. Fund for Animals (U.S.A).

Supportive of Wolf Control (Poisoning)

1. Sinkut Mountain Cattlemen's Association
2. Cluclutz Livestock Association
3. Fort Fraser Livestock Association
4. Nicola Stock Breeders Association
5. North Okanagan Livestock Association
6. Peace River Regional Cattlemen's Association
7. Smithers Predator Control Committee
8. B.C. Cattlemen's Association
9. Bulkley Valley Cattlemen's Association
10. Cariboo Cattlemen's Association
11. Skeena Guides Association
12. Nanaimo Fish and Game Club
13. Nadina Rod and Gun Club

Questioned need to control wolves to protect caribou

1. B.C. Society for Prevention of Cruelty to Animals (SPCA)
2. Humane Educators of B.C. (Kindness Club Affiliate)
3. Shuswap Naturalists
4. North Okanagan Naturalists Club
5. Whiterock and Surrey Naturalists
6. Arrowsmith Natural History Society
7. Penticton Naturalists Club
8. Telkwa Foundation
9. Aloutette Field Naturalists
10. Canadian Nature Federation

11. Northland Bowhunters
12. Cowichan Valley Natural History Society

Petitions opposing the use of poison
1. Pitt Meadows – 44 names
2. North Vancouver – 36 names
3. Victoria Wolf Cub Pack and Brownies – 36 names
4. Westbank, B.C. – 20 names
5. SPCA Kelowna – 59 names
6. Victoria – 25 names
7. Edmonton, Alberta – 33 names
8. Brentwood Bay – 530 names

Petitions supporting wolf poisoning
1. Takysie Lake, B.C. – several names on a letter

Following are the main points of concern expressed in the letters and the percentage of times mentioned.

		Percentage
1.	Opposed to use of poison	11
2.	Criticized use of 1080 poison	21
3.	Poison cruel or inhumane	33
4.	Other animals are killed from eating poisoned bait	17
5.	Poison interferes with the balance of nature	20
6.	Poison has a chain reaction	6
7.	Object to spending tax money to poison wolves	7
8.	Object to extermination of wolves	8
9.	Wolf control unnecessary	12
10.	Wolves sociable, friendly animals	7
11.	Do not know the facts	8

Alternative solutions offered

Twenty-six percent of the people offered one or more solutions to the poisoning of wolves. These are:

		Percentage
1.	Protect cattle some other way	14

27

2. Review wolf problem using an independent scientist 4
3. Reimburse ranchers for losses 3
4. Let the ranchers protect themselves 3
5. Close or control hunting 6
6. Spend money allocated for wolf control on research 2

Discussion

The press first drew attention to the plan to control wolves in the Bulkley Valley region; then there was the account of a research proposal from Dr. John Elliott, regional wildlife biologist in Fort St. John involving Level Mountain and the Horse Ranch in northern B.C., and finally a separate proposal to eliminate wolves in a ranching area on the Bonaparte Plateau to protect livestock. The newspaper headlines along with some of the erroneous or misleading excerpts from these are included in this report (see following pages 29-32).

Mr. George Clement's (Association for the Protection of Fur Bearing Animals) description of the article of October 2 and October 5, 1978 gave a detailed account of plans to poison wolves using 1080 poison and the wide coverage this received was undoubtedly responsible for the high proportion of comments (33 percent) dealing with cruelty. Moira Farrow's articles in the Vancouver Sun dated October 2 and October 5, 1978 gave a detailed account of plans to poison wolves in the Skeena and Omineca regions. She referred to a secret memorandum, inadequate information, and opposition by biologists. It would seem that these articles were designed to elicit a public response in opposition to wolf control.

There was surprisingly little evidence at the time of large scale individual solicitation of letters. In fact, only in Surrey and Maple Ridge was there any suggestion of an organized letter writing campaign.

The telegrams from the various livestock associations may have been solicited by the parent body, the B.C. Cattlemen's Association. This would be expected in view of a newspaper article requesting that people write to the Premier of B.C. in opposition to wolf poisoning.

Three percent of the letters made reference to Farley Mowat's book, "*Never Cry Wolf.*" Notwithstanding this small percentage there were indications that many people had been influenced by this book.

Farley Mowat himself wrote to say that he had been deluged with mail. Obviously he had become an authority in the public mind. Mowat advised the Minister that a film was being made, based on his book.

Only a very rudimentary understanding of animal ecology is evident on part of the people objecting to either wolf poisoning or wolf control. The response was largely an emotional one. Surprisingly, not a single writer expressed any empathy whatsoever for the animals that fall prey to wolves. This was accepted as a part of nature and the belief that man has no right to interfere. Man himself, was not recognized as a part of nature.

It was commonly believed that wolves prey only on the old and infirm components of ungulate populations. One writer even suggested the benefits of wolf predation to the cattle industry in helping to keep livestock healthy.

One cannot assume that the writers represented the true level of public understanding of wildlife matters. However, it is evident that the Wildlife Branch must do much more than at present to educate the public about wildlife matters, particularly the elements of animal population dynamics and the benefits of management.

Socred's wolf campaign branded as savage

"... The wolf was put on the protected animals list until a few years ago when the present wolf-hating Socred Government took it off. Perhaps the wolves are also killing a few of the caribou that the Socreds encourage wealthy American "sports" hunters to come up and kill and wound and maim in B.C. and that would never do!"

Poison use Political
Victoria Times, December 11, 1978

"A leading wildlife biologist (Dr. Ian McTaggart Cowan) said a better idea would be to identify the particular animal responsible and have an experienced hunter shoot it."

Wolf poisoning 'act of despair'
Stephen Hume, *Victoria Colonist* about December 11, 1978

"McTaggart Cowan said 1080 was a slow killer, it may take a couple of days."

Man Kills for Sport
Victoria Colonist, December 12, 1978

"...with anger and disgust I read of the governments proposed inhumane killing of our few remaining wolves in B.C. to pacify a handful of ranchers in the Kamloops area."

Torture in the name of necessity
Letter to the *Victoria Colonist,* December 6, 1978

"These animals (wolves) have their place in nature and their threat to the balance of species is pure propaganda designed to serve commercial interests as every unbiased naturalist knows." ... (wolf control) ... could be done by allowing the ranchers themselves to shoot wolves which are harassing their stock."

Horror Death Program in B.C.

"It will be interesting to see if enough B.C. voters will pressure their M.L.A.'s to stop this horror program before our wolf population has been wiped out (as it was 20 years ago by another government) to the point where the wolf was put on the protected species list."

Sierra Club on side of wolves

"... control should be limited to problem animals when the Fish and Wildlife Branch is satisfied that the rancher concerned has taken all responsible action to minimize predation."

Sierra Club lobbies against wolf poisoning

"... predator control in order to provide a larger 'surplus of caribou for hunters ... constitute(s) a subsidy to caribou hunters by the taxpayer."

Wolf poisoning cruellest of deaths
Victoria Colonist, December 2, 1978.

"1080 ... ensures him (wolf) and other creatures in contact, the cruellest of lingering deaths."

Saner methods than poisoning
Victoria Colonist, Dec. 5, 1978.

"As anyone knows, the real damage to game animals is done by humans, not wolves. Wolves actually benefit the game by culling the weak and unfit. Nature always keeps its own unless man interferes. There are saner ways ... using dogs to guard the herd (or better yet, a person), letting the cows grow horns or shooting individual wolves known to be calf killers are just a few."

Wildlife at wolves door
Colonist, December 8, 1978.

Article by Stephen Hume quoting Dr. Patrick Moore of Greenpeace Foundation.

"(Lithium chloride) treated carcasses taught coyotes not to eat sheep ... there was no reason why it wouldn't work for wolves."

Ranchers favour war
Colonist, November 25, 1978. Article by Stephen Hume.

Stephen Hume quotes a veterinarian and drug committee chairman of the B.C. Veterinary Medical Association to the effect that wolves control their own birth rate – "they limit it to the area they are in ... I can't believe half the complaints about wolves are related to wolves at all. We drive wolves to this by destroying their habitat ... The fact is they may have taken the weakest animal that wouldn't have survived anyway. There always has to be a fall guy."

Fiction writers should not be unchallenged experts in the public mind. Branch biologists should be allowed and encouraged to advise the press and the public about the importance of wolf control in game management.

The newspaper articles during 1978 were undoubtedly the main source of information and inspiration for letters to the Premier and the Minister of Environment. Many of the headlines were provocative as noted from the examples illustrated. Most of the public concerns about the proposed resumption of wolf control expressed in letters to the Minister and the Premier were to be found in articles and letters in the newspapers. There was also evidence of bias on the part of some newspaper reporters. Many letters contained inaccuracies and ignorance of factual information.

It is unfortunate and harmful to wildlife resources and public benefits when negative letters to newspapers are able to have such a detrimental influence on proposed science based management programs.

7

The Muskwa and Kechika Wolf Management Research Project – 1983

Alaska and the Yukon were the first to experiment with a new method to achieve site-specific wolf control by shooting from helicopters. It was even more humane and more selective than 1080 and its effectiveness was demonstrated in significant increases in moose and caribou calf survival in study areas in Alaska and the Yukon.

Dr. John Elliott, the regional wildlife biologist in Fort St. John, was eager to demonstrate that aerial wolf control could be undertaken successfully in northern B.C. He proposed a research project in 1981, duplicating those in Alaska and the Yukon by comparing moose survival between areas with no control of wolves and adjacent ones from which wolves were removed.

The operation, referred to as the Muskwa and Kechika wolf management research project, was designed to parallel the Alaska and Yukon studies that had for the most part escaped attention from animal rights groups in the south.

When the project finally got underway in 1983 there was an incredible outburst of public opposition directed at the Provincial Government and the Fish and Wildlife Branch, carried mainly by Victoria

33

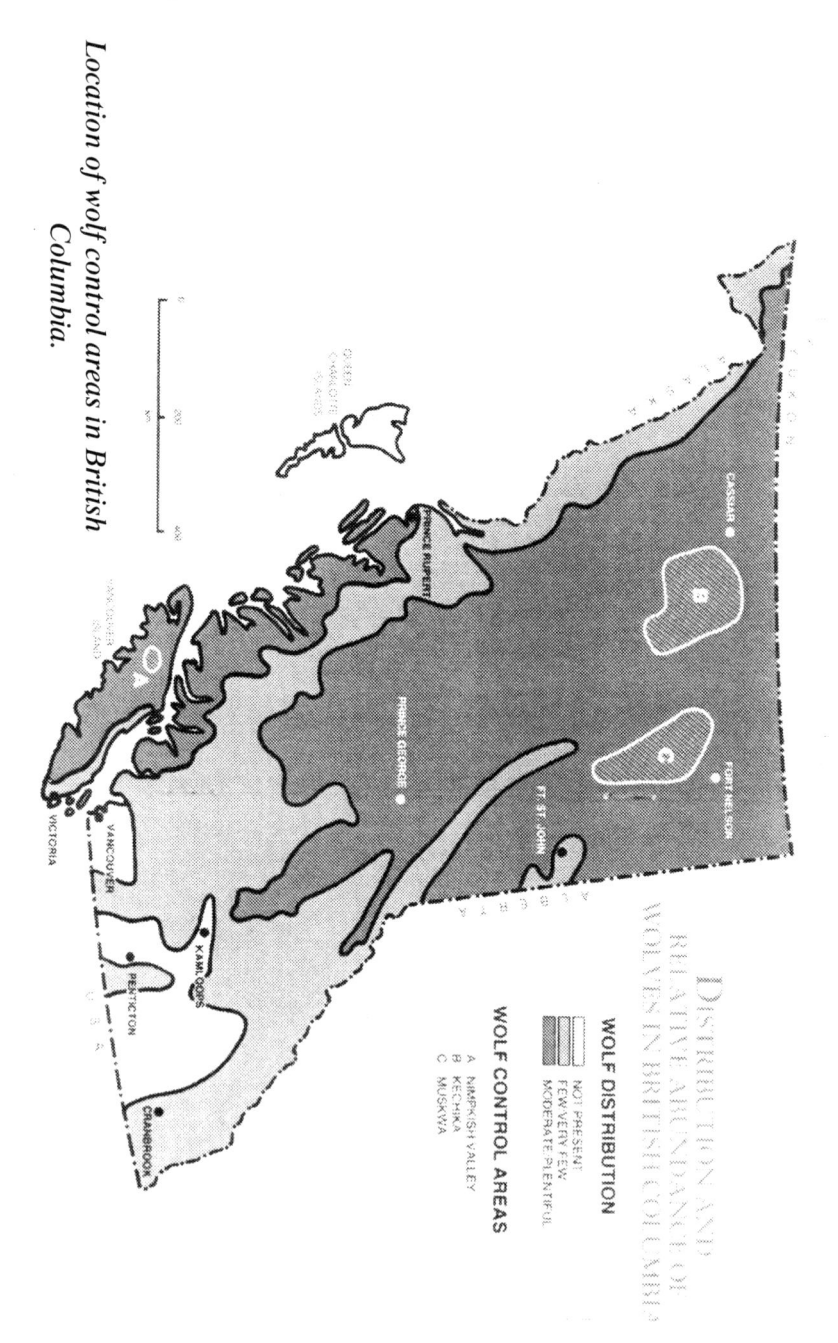

Location of wolf control areas in British Columbia.

and Vancouver newspapers.

Vociferous criticism by animal rights proponents and others had conditioned many people in 1978 to think of the wolf as an icon. As a result, the aerial control research project was terminated by the Provincial Government in the interest of political correctness. The issue went to court and a technical error was found in the Wildlife Branch not having the necessary permit to make shooting from aircraft legal. All was not lost, however, as Dr. Elliot had collected sufficient data in time to support the conclusions of the Alaska and Yukon biologists.

On Feb. 21, 1983, Jack Ramsay, the *Vancouver Sun's* correspondence editor reported an organized letter campaign in which 90 percent of the letters received were opposed to wolf control in the Muskwa. He said, "I have absolutely no doubt that the American letters are orchestrated by, among others, the Animal Protection Institute of America in Sacramento, California. Many of the letters contain the same language as in the institutes 'Emergency Alert'."

Project Wolf USA, which organized the boycott in Seattle, collected a 1,700-name petition on city streets from people who wanted the hunt stopped.

On Feb. 28, 1984, Ramsay wrote, "Since the story broke on Jan. 11, the Sun has received more than 500 letters on the issue, of which 35 have been published. More than 90 percent deplore killing the wolves." Thirty-eight percent or 190 letters came from British Columbians. Letters from the other provinces included:

Ontario	26 letters
Quebec	10 letters
Alberta	6 letters
Manitoba	2 letters
Saskatchewan	1 letter

The Federal Environment Minister, Charles Caccia stated that "the issue was of great concern to the Canadian public."

Orchestrated letters received from 37 states and the District of Columbia were as follows: California – 51, New York – 23, Washington State – 18, Pennsylvania – 11, Texas – 11, Illinois – 8, Michigan – 7,

Wolves and People

Wolf control program dogged by pure myth *Times-Col. Feb 3/84*

Leave the north to the northerners *Sun Mar 1-84*

Editor, The Sun, Sir: *Feb 11/84*

Pitiful ignorance the worst feature of wolf argument

People or wolves, where's priority? *Sun Jan 30*

Men, not wolves, are vicious killers *Sun. Van. 17/84*

B.C. wolf kill called 'massacre' *Sun. Jan 28*

Wolf-kill disrupters get help from Indians *Colonist Feb. 13/84*

Wolf kill proceeds despite protest *Sun Feb 9/84*
By MOIRA FARROW
Sun Staff Reporter

'There's no justification' for B.C.'s wolf control *T-C Feb 22/84*

A 10 The Sun WED., FEBRUARY 15, 1984

Wolf kill fighters lose battle, start war
By MOIRA FARROW

Newspaper headlines Muskwa 1984

The Management Imperative and Mythology of Animal Rights

Wolf-kill plan draws howls
By MOIRA FARROW *Colonist Jan 14/84*

'Politicians more vicious than wolves' – Mowat *TC Feb 28/84*

Wolves no problem, just bipeds: Caccia

WOLVES RIP NDP APART *Prov Jan 31/84* *Sun Jan. 13/84*

FARLEY MOWAT ...politicians more vicious

Americans financing wolf kill *Sun Feb. 2*

Emotional letters are short on facts *FEB 28 SUN*

GREEN PARTY ALSO ANTI WOLF HUNT *Prov. Feb. 27/84*

By MOIRA FARROW 'Why death sentence for the wolf?' *Jan. 19/84 Prov.*

Trap, poison, kill — some solutions! *Province, May 14/86*

U.S. travel agents boycott B.C. to fight wolf slaughter *Sun Feb. 6/84*

Brummet rejects bid to halt B.C. wolf kill *SUN MAR 6/84*

New Jersey – 7, Wisconsin – 7, Florida – 5, Maryland – 5, Massachusetts – 5, Missouri – 5, Arizona – 4, Colorado – 4, Connecticut – 4, Ohio – 4, Oregon – 4, Iowa – 3, North Carolina – 3, Virginia – 3, Alaska – 2, Washington D.C. – 2, Georgia – 2, Louisiana – 2, Wyoming – 2, and one letter each from Delaware, Idaho, Kansas, Maine, Minnesota, Mississippi, Nevada, New Mexico, South Carolina, Utah and West Virginia.

It was a mistake for the Provincial Government to have capitulated to the organized letter campaign even though aerial shooting from a helicopter had not been covered by an appropriate permit. The antis won a victory and unfortunately learned that letter campaigns to newspapers are effective in obstructing wildlife management programs they oppose.

Paul Watson in search of Dr. John Elliot

Staff Reporter

"Paul Watson dons cross country skis today in pursuit of the northern B.C. wolf hunt, but it seems he's on the wrong track.

In fact, he's off in the opposite direction to his quarry – B.C. government wolf hunter and biologist John Elliott – according to an environment ministry spokesman in Prince George.

"They're headed for the wrong range, which is really kind of funny and sad," chuckled Don Morberg, who said Elliott has killed 64 wolves in the Muskwa region towards a goal of 330.

In his second anti-hunt campaign in the last month, Watson is setting out form Prophet River, 100 kilometres south of Fort Nelson.

He believes Elliott is based at a ranch about 65 kilometres west of there.

But Morberg says Elliott is based at a ranch about 150 kilometres northwest of Watson and more than 100 kilometres west of Fort Nelson.

Letter to BCWF from John Elliott, December 1, 1983

December 1st, 1983
B.C. Wildlife Federation
5659 – 176th Street
Surrey, British Columbia
V3S 4C5
Dear Sir:
Re: Muskwa Project

I would like to thank the B.C. Wildlife Federation, it's Directors and membership, for it's support of the Muskwa Project. I anticipate that the wildlife in the Muskwa area can surprisingly readily be increased to 12,000 elk (from 4,000), 6,000 sheep (from 3,000), 20,000 moose (from 12,000), 1,000 caribou (from 500), plus goodly numbers of deer and mountain goat. These numbers for vigorous populations should support a hunter harvest in excess of 7,000 per year.

The concept of managing predator numbers as a tool in achieving increased numbers of hoofed big game runs counter to much of the conventional biological dogma. In consequence the wolf work which has been undertaken in the northeast of British Columbia has been scrutinized very closely. Because 'umpteen' years and millions of dollars have not been expended studying the matter, every detail of the wolf-ungulate relationship has not been minutely examined.

However, let us consider the information which led to the Muskwa project being proposed.

1. Moose in the northeast appear to have declined by up to 100,000 animals to less than 50,000 since 1975.
2. Mountain sheep in the northeast appear to have declined by about 9,000 animals to less than 7,000 since 1975.
3. Elk in the northeast appear to have declined by up to 1000 animals to less than 4,500 since 1975. Many of the smaller herds have virtually disappeared.
4. Caribou in the northeast appear to have declined by about 7,500 animals to about 2,500 during the 1970's.
5. Examining a number of areas in the north, it was found that as wolf numbers increased from 1 per 130 square miles to 1 per 26 square miles, caribou calf proportions declined from 27% to 12%.
6. When wolves were reduced on the Horseranch Mountain Range, calf caribou proportions quadrupled for the three years of reduction. The herd increased by 42%. A nearby herd declined 43% during the same period. When the wolf reduction stopped, calf proportions plummeted and the herd declined within three years to less animals that were present at the start of the program.
7. Examining a number of areas in the north it was found that as wolf numbers increased from 1 per 100 square miles to 1 per 18 square miles, moose calf proportions

declined from 28% to 10%.
8. In the Kechika it was estimated that the moose were declining by 50% every three years. In a small portion where wolf numbers were reduced, the herd is now estimated to double within four to five years. Indeed, while normal fall yearling to adult bull proportions are less than 10%, the low wolf area in the Kechika the second fall after wolf reduction showed one third of all bulls to be yearlings. Calf proportions were three times those in the normal wolf area.
9. Following wolf reduction mountain sheep lamb proportions doubled and yearling proportions tripled over those in normal wolf areas.
10. A low wolf elk area showed twice as many calves and three times as many yearlings as the normal high wolf areas.
11. There are up to 4,000 wolves in the northeast. Utilizing standard food consumption rates and estimating average prey weights (remember, many are juveniles), wolves in the northeast appear to consume in excess of 50,000 hoofed big game animals per year.

While details of sample size, methodology, or other aspects are always open to criticism, it is rather suspicious that game declines in the northeast are extreme, that wolves are abundant, and that the wolf has played a role in the declines of game.

TONY BRUMMET:
"Predator control will always be a part of wildlife management. If we let natural cycles take their own course, then let's get out of wildlife management."

In Wildlife Review magazine the Hon. Tony Brummet from rural northern B.C. spoke out in support of the Muskwa-Kechika wolf management study, March 1984.

Wolves and People

It is encouraging that when the hunters cry wolf they are prepared to dig into their pocket and do something about it.
Yours truly
J.P. Elliott
Regional Wildlife Biologist
Omineca Peace Region.

(The BC Wildlife Federation raised $34,000 in a lottery in 1983-84 to contribute to the Muskwa project).

Article from *Wildlife Review magazine*, March 1984

Brummet, who was appointed environment minister on May 26, 1983, is a seasoned outdoorsman and self-described casual hunter who continues to spend much of his vacation time in the bush. Since he moved to Fort St. John as a school principal in 1965 he has witnessed the decline of game populations in the backwoods he travels.

"I support the predator-control program because I happen to believe in it from my own experience and from what the evidence shows me. I think it's essential," Brummet says. "To say that we, the politicians, designed the program for some political reason is sheer idiocy."

Like others in his ministry, Brummet was surprised that the wolf issue was raised in 1984, three years after the projects in the Muskwa and Kechika first began.

"All that happened, you see, is that Paul Watson jumped on it and made it an issue as though we'd just instituted wolf-control programs. It's been going on in Alaska, it's been going on in the Yukon, it's been going on in Alberta; it's been going on in B.C. Predator control is not new: the issue, the public focus on the issue is what happened this year."

Brummet says he refused to meet with Watson because Watson had spent considerable time publicly attacking the wolf-control programs. Any meeting would have been a waste of time, he says. It appeared obvious to Brummet that the plan was to use the minister's office as the setting for a publicity stunt "to have the cameras there to call me names. I can do a lot more productive things than that."

The minister also refutes the suggestion by George Clements that he is unfit to preside over the environment ministry because of his feeling that the wolf "is one of the most dangerous, vicious, wasteful and unrelenting killers in existence."

"That's what I said, and if you take that and look at it, is there anything in that statement that is untrue? Maybe it was bad politics to make that statement, but I guess maybe I'm not a politician primarily. Of that statement there's not one point that I'd need to retract on the basis of its merits."

Brummet says there are territories in B.C. which can sustain far more big game animals and the commitment of his ministry is to provide more wildlife, whether it's for viewing, hunting, aesthetic or other reasons.

"Predator control will always be a part of wildlife management," Brummet says. "If we let natural cycles take their own course then let's get out of wildlife management. Natural cycles will work if you leave them alone, if you don't care, if it doesn't matter. But there was a conscious decision made that we should get into wildlife management. Wildlife management then leads you to enhancement programs, to limiting access, to predator control. The basic thinking is that you either accept wildlife management, with all the things that the technical people come up with from the evidence, or you abandon wildlife management and leave it to nature."

But will BC's wildlife ultimately be better off if its welfare is left entirely up to nature? Even in a province like British Columbia, which promotes a wilderness image to the world, there is little, if any, true wilderness left. We have upset the balance just by being here, and many conservationists feel it is now our responsibility to restore it. With wildlife we can do more than merely replenish what we've lost:

Wolves and People

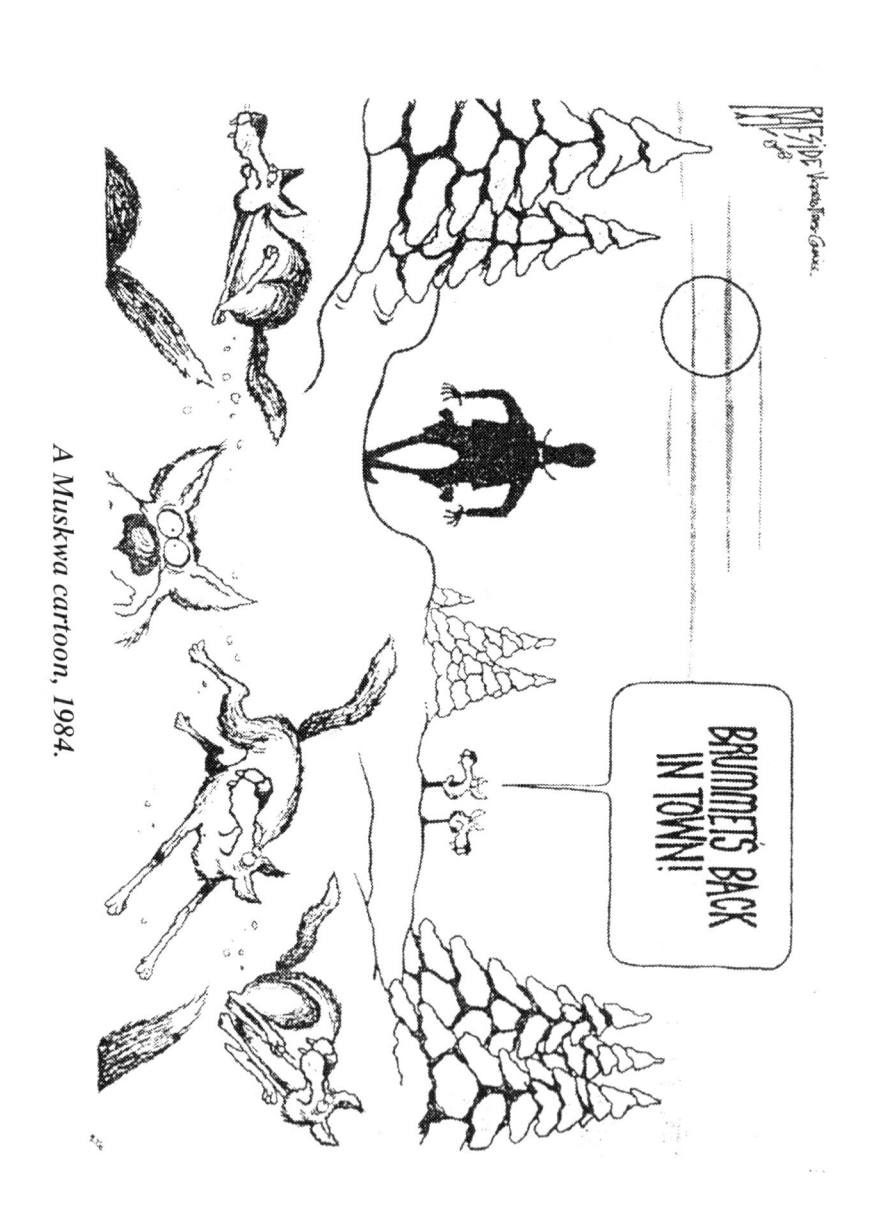

A *Muskwa cartoon, 1984.*

it is possible in some cases to improve upon natural habitats and their ability to support wildlife populations. Our woods and wetlands, our streams and grasslands could be inhabited by more and healthier wild animals through management techniques like habitat enhancement. If this is interference with nature, then we may be wise to interfere judiciously.

The question in the predator control debate is the degree of management. Should we continue to use public money to sustain our $100-million hunting industry by insuring a supply of big game or should we limit wildlife management activities to those which respond only to non-consumptive use demand, substantially reducing hunting and fishing opportunities? Economically, there is endless justification to support recreational hunting and fishing in B.C. – and its popularity attests to the demand of the market. But we've learned from the wolf controversy that an enormous number of people in this province are adamantly opposed to killing wildlife.

Muskwa cartoon, 1984.

Wolves and People

If the wolf-kill program in northeastern B.C. can be truly justified from a wildlife management point of view – and divided professional opinion makes this far from clear – then support for, or opposition to, the program remains a matter of personal conscience. The nature of this issue itself inevitably polarizes opinion, because matters of morality have no middle ground. And so, the controversy will continue, because it represents a test by which each of us defines ourselves as inhabitants of our environment.

In January 1984, this writer wrote as follows to the Honourable Anthony Brummet, the newly appointed Minister of Environment. I write to you, not only as a director of the B.C. Wildlife Federation, but also as a former director of the Fish and Wildlife Branch.

You have inherited a ten-year hiatus of almost total government indifference to the need to keep the public informed on game management issues. There has been a succession of senior people in the Ministry of Environment who have been reluctant to permit, or who have disallowed, the dissemination of factual and educational material that might arouse controversy. Education is essential to help the public to understand issues such as predator control and the economics and necessity of hunting.

I have complete confidence, as a hunter and game manager, in the advice you will have received from Dr. John Elliot; and I respect the concerns of guides, hunters, and other northern people due to the renewed unnecessary destruction of ungulate populations by wolves. It will be a lasting tragedy if you are forced by political expediency to curtail Dr. Elliot's program. We cannot afford such triumphs of ignorance over logic.

I trust that you will seek out the sources of the communications problem within your Ministry and rectify them. If this is not done, I predict that the wolf issue will not be the only one to fall on your shoulders and that our province will measurably suffer from a decline in the economic and social importance of hunting.

In an earlier personal interview when I was no longer director of the Fish and Wildlife Branch I warned the minister to beware of biased advice from unsympathetic senior public servants in his Ministry. I recall a remark he made when I mentioned the inclination of some

The Management Imperative and Mythology of Animal Rights

less objective senior staff to tell their Minister and deputies what they thought they would like to hear. Brummet indicated that he had become aware of that problem and wanted no advice from "yes" people in his Ministry.

Had it not been for Tony Brummet, Dr. Elliot's Muskwa project would have been terminated even sooner.

In my experience as a public servant, the most progress in game management has occurred when either the Premier or the Minister in Office was a hunter and fisherman. The Hon. Ken Kiernan, however, told his critics to talk to the biologists, not the politicians, if they wanted good game management. "What do you want" he asked them, "management by a bunch of politicians or by professional biologists."

On another occasion, during the Muskwa research controversy, Premier Bill Vanderzahm is alleged to have told his critics that British Columbia did not need outside advice on how to manage its wildlife.

By 1997, after nearly 20 years of minimal predator control by hunting and trapping in the East Kootenays, wolves increased and were affecting deer, elk and moose populations in parts of this important region for big game hunting.

8

A Report from the Guide Outfitters in British Columbia – Wolf Predation in the 1990's

At the turn of the century professional big game hunters (guides) were the first to bring world wide recognition to superlative hunting success in British Columbia's northern wilderness. As early as 1905, the Provincial Game and Forest Warden, A. Bryan Williams recognized the important role guides had in attracting significant foreign capital that contributed to the economic growth of the province. They were also the ones who pressured the Provincial Government to protect southern B.C.'s game from American hunters illegally crossing the border.

Due to the continued concern in the 90's over wolf predation in northern areas of the province, Dale Drown, Executive Director of the Guide Outfitters Association of B.C. mailed my wolf questionnaire[1] to northern guides to help the writer review wolf predation as if affects big game hunting and ungulate conservation.

---

[1] Please see Appendix 1 for copy of questionnaire sent to guide-outfitters in northern British Columbia.

48

The Management Imperative and Mythology of Animal Rights

Thirty-six replies come back from guides whose territories have wolves. Almost 100 percent (35) indicated their concern about predation. This represented the accumulated experience of 882 hunting seasons or average individual involvement of 24.5 seasons of big game guiding.

When asked how many domestic animals had been killed in their area by wolves during their years guiding, 26 guides gave the following information:

Horses – 69
Cattle – 33
Dogs – 27

Table 1. Order of big game species listed by the average percentage of overall occurrence in harvest and overall economic values (income).

Occurrence	Income
1. Moose: 88%[2]	1. Moose: 81%[3]
2. Mountain Goat: 61%	2. Mountain Goat: 59%
3. Caribou and Black Bear: 44%	3. Black Bear: 47%
4. Mountain Sheep: 33%	4. Grizzly Bear: 44%
5. Rocky Mountain Elk: 28%	5. Mountain Sheep: 42%
6. Grizzly Bear: 25%	6. Rocky Mountain Elk: 28%
7. Mule Deer and Wolf: 22%	7. Caribou: 39%
8. Cougar and White-tailed Deer: 3%	8. Mule Deer: 19%
	9. Wolf: 6%
	10 Cougar and White-tailed Deer: 6%

---

[2] Moose occurred in 88% of all guiding areas reporting
[3] 81% of guiding income came from moose hunting

Wolves and People

A single wolf attacked and carried off a yearling stone sheep from a salt lick beside the Alaska highway (photos taken through a vehicle window).

Tim Mervyn Crew photos via Reg Collingwood

A horse that survived wolf attacks (Guy Anttila photo).

A Norwegian elk hound killed and partially eaten by two wolves on a snow covered lake. It went down seven times before dying (Milt Warren, photo).

Wolves and People

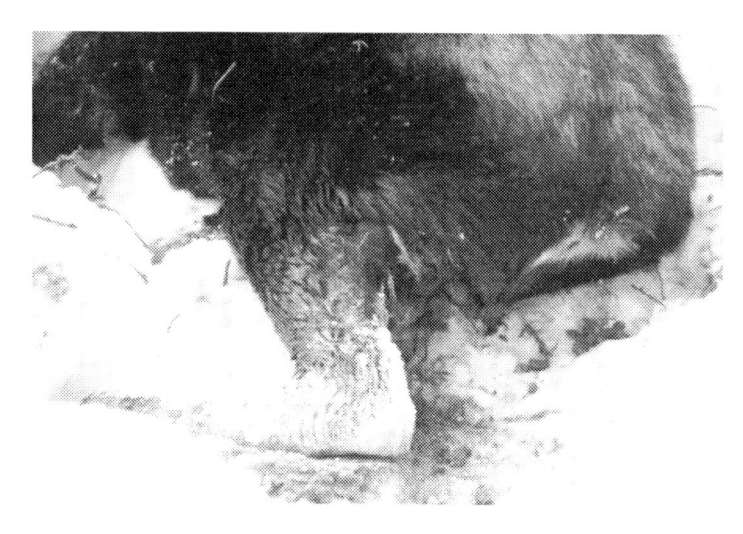

Hamstrung moose killed by a wolf pack (Reg Collingwood, photo).

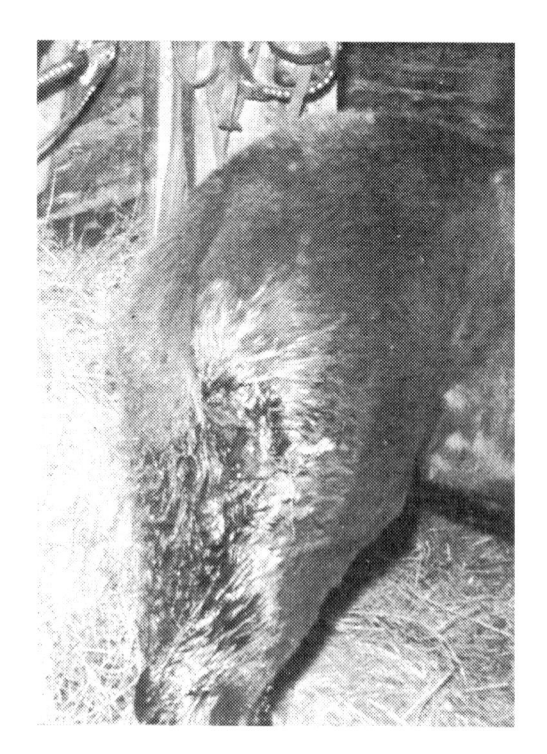

Hamstrung colt (Nancy Ball, photo).

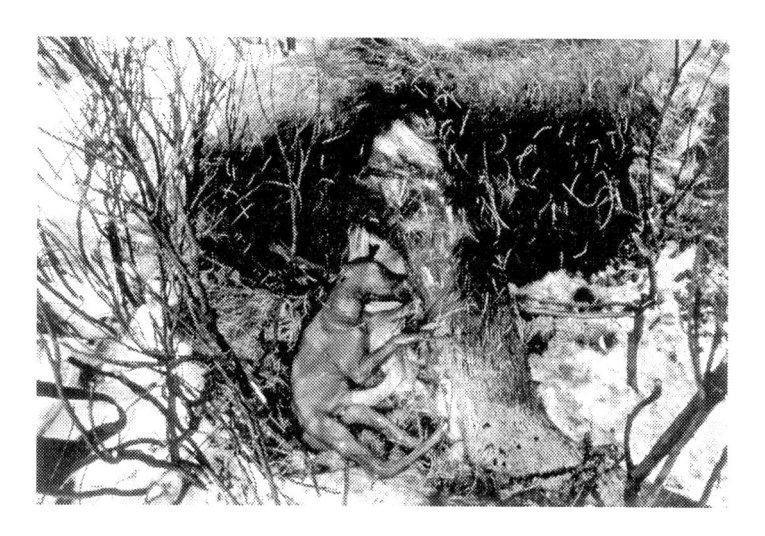

Hamstrung pregnant cow moose. Fetus removed to show development (Reg Collingwood, photo).

Crissey, the author's springer spaniel attacked by an 85 lb. wolf on Sidney Island, near Victoria, 1986 (please see Politically Incorrect, page 136).

Wolves and People

It is apparent that the order of species occurrence in harvest is very similar to the species economic importance to the guides.

Table 2. Reported trend in wolf abundance.

Wolf Population Trend prior to 1990	Wolf Population Trend since 1990 (after Wolf Control)
19 guides reported wolves increasing	21 guides reported wolves increasing
12 guides reported wolves fluctuating	7 guides reported wolves fluctuating
4 guides reported wolves stable	5 guides reported wolves stable
0 guides reported wolves decreasing	1 guide reported wolves decreasing
Total guides reporting = 36	Total guides reporting = 34

Thirteen guides reported their highest personal count of wolf pack size from the air as follows: 15, 19, 14, 37, 15, 25-30, 10, 24, 18+, 12, 22, 20 and 44 for an average observed maximum pack size of 21.

Thirty-one guides reported the largest pack size as reported from someone else. The numbers are as follows: 12, 20, 15, 20, 30, 27, 90+, 25,12, 8-10, 7, 24, 15, 17, 14, 72, 50, 12, 35-40, 16, 14, 11, 31, 18, 12, 21, 13, 12, 22, 25 and 25 for an average maximum observed pack size of 21.

Thirty-one guides responded to the question whether or not wolves were observed to deplete furbearer populations. Thirteen replied yes, 7 replied no, and 11 didn't know.

Five guides knew of wolf attacks on people. Twenty-five respondents knew of no such attacks.

Ten guides reported knowing of a wolf pack turning on a wounded member of the pack. Nineteen respondents had no such knowledge.

Sixteen guides reported what appear to have been hydatid cysts in big game animals caused by the granular tapeworm, which is carried by wolves and can also infect people. No one knew of people infected with hydatid disease.

Twenty-seven guide outfitters knew of the book "Never Cry Wolf"

The Management Imperative and Mythology of Animal Rights

by Farley Mowat. Six had not heard of it. Fourteen said they knew that high school students were required to read the book listed as non-fiction.

Permission was given by all guides to quote them and report information provided.

Following are the voluntary comments from 25 Guide Outfitters.

Bob Nielsen – Tweedsmuir Park (29 years guiding)

Wolves fluctuate with the game and also the winters. I have found they prey on caribou early, and mostly bull caribou. I have observed a lone wolf taking a calf moose in mid June. They leave the Park in late November, some to the east and the others north to Ootsa and Nadina. They have been very hard on our recovering deer population. We had a very healthy population coming back in the Kenny dam watershed, both mule and white-tailed deer. A season was opened for whitetails a few years back. Now, with a couple of harsh winters and predation they are extremely low in numbers again. We observed 5 deer kills and one moose kill in a 5 mile stretch of shoreline.

For some reason, last fall (98) in North Tweedsmuir (our hunting area) we saw the least signs of wolf since 1981. Reason? I'm not sure, but the following year we shot a couple that had mange really bad. Along with a very mild winter with little snow, I believe they had tough hunting.

I'm sure you remember when we had some predator control in the 50's, 60's and early 70's when Milt Warren was active. In the early 60's when Northern Mountain Airlines was first established in Prince George and Fort St. James (my father founded the airlines), float and ski planes were opening up the north. Wolves were very abundant but many were harvested by 1080 etc. via aircraft and hunting. I can't remember the numbers but recall when Milt and N.M.A. baited Stuart, Trembleur and Nation lakes and several days later picked up a large number of wolves. From that time on the moose population recovered to its highest ever. Outfitters, like the Davidson's were known to take 50 moose or more in one season off two lakes. We used to be able to harvest our moose within a 15 mile radius of town. Now you have a better chance of getting a wolf.

Leonard Ellis – Bella Coola, B.C. (20 years guiding)

Lived in Ocean Falls for 12 years. Lost my 6 year old male black Labrador to a pack of wolves. Lost 8 male dogs in Ocean Falls in 10 years. The natives say the wolves send a female wolf in heat close to town to lure the male dogs out. It seems to be true. I found a young wolf eating the hind quarters off of a young doe blacktailed deer in Gunboat Passage one day. The deer was healthy and the wolf was eating it alive. It would have been good to have got that on film.

John Robidon – Fort Nelson, BC. Dunedin River Outfitters (4 years guiding)

Concerned about wolf predation – reducing all antlered game. Found kills of trophy class animals at all times of the year. We need wolf control badly.

Albin Hochsteiner – Osoyoos, B.C. (35 years guiding)

We have no wolves. From everything I have heard, I hope we never do have any. I am a hound hunter first and foremost and I know from people who have first hand knowledge that wolves are dynamite in relation to dogs. If you use hounds in wolf country, you better figure on losing some to these predators.

Ray Collingwood – Smithers, B.C. (30 years guiding)

I have been concerned (about wolf predation) ever since we started outfitting in 1969. The great wolf predation started in Spatsizi about 1974/75, peaked in 79/80 and by 82/83 the moose population had dropped to around 600 from some 3,000. Caribou not as drastic, 1200 maximum from 3500. In more recent years, particularly 94 and 95, the

Stone Sheep population on the Spatsizi Plateau (200 – 220) crashed drastically. We estimate about 35 sheep remain. It's strictly predation. There is no lungworm disease. I know the packs, their colour and territory. I might add the same Plateau area had 65-70 goats from the 50's to 75/76. About 15 survived to date.

In regard to my experiences with wounded wolves while winter wolf hunting. I've tracked in the snow a wounded wolf for 2 days. The rest of the pack seemed to hold up and assist as their pack members lay down and rested. I distinctly recall this behaviour on 4 occasions.

In respect to the meeting of packs, I witnessed at the Ross and Spatsizi juncture in 1991, a lead wolf who met a pack of 5. The lead (alpha) wolf ventured out on the ice where 2 from the pack of 5 challenged him. There was a fierce fight and within 2 minutes the lead wolf had killed one of the pack in 30-40 seconds. His partner limped away with wounds that looked to me to be too severe to survive. I've witnessed a single wolf kill a 3 or 4 year old bull moose.

(In 1975 the author found the skeletal remains of a 5 year old Stone ram in wide open country in Spatsizi. In the absence of escape terrain, it may have been overtaken by wolves. It did not look like sheep terrain where a guide would have taken his hunter and besides it would not have been a legal ram. Not too distant but in typical goat country I found the weathered remains of a female goat. A pack of 13 wolves, described in my book *Politically Incorrect*, were also observed in the area).

Jerry Geraci – Upper Stikine River Adventures (14 years guiding)

Wolf predation in my area is keeping the numbers of game down instead of letting its numbers increase. Lamb sheep down in 1996, 97 and 98. Out of 7-8 ewes you will see 1 to 2 lambs. Not even half. Same with cow caribou.

We have found trophy class moose and caribou taken down by wolves. Most recently (fall of 97) on the horse trail wolves had taken a 5 by 6 bull caribou, about 9 years old. I watched wolves trap a cow caribou in the Stikine River. Guts were visible and she was hamstrung. A single wolf took down a bull moose in March.

We have shot into a pack of 11 wolves and the hunter wounded a pup. The others turned on the wounded one. I know of two such occurrences.

I am not against wolves. They have their place but if we are going to manage all other game, then they should be managed as well. The only reason they are not is political, not biological.

Terry Stocks – Fort St. James, Kamloops (17 years guiding in BC)

Two beta female wolves wounded, one in February and one in March were turned upon by the pack. Wolves have been returning to the southern areas of the Province where they were reduced in past years i.e. 98-99. Wolves are preying on the Kamloops Lake sheep herd.

Bill McKenzie – Gana River Outfitters (26 years guiding in BC, 15 years in NWT)

Prior to going to the NWT, the previous owners of the guiding territory tried wintering horses in the area with poor results, due to wolf predation. It was tried for two winters with over 50% loss to wolves. This was with people living on site all winter and patrolling and feeding daily.

I have personally had several incidents where I was worried by wolves. Usually when I did not have a gun handy and they seemed to know there was little danger from me. Twice I have met wolves on a trail up close with no gun, and been snarled at. I do not fear them as a rule, but I do not trust them and I feel that there is a real danger if young children are vulnerable. I have seen a man who was attacked by a year old wolf while feeding it in a pen. He was badly mauled in about 10 seconds and was lucky to have fallen backwards through the wire when attacked, or I am sure the wolf would have killed him within another 30 seconds.

My crew reports at least 20 adult bull caribou, 6 to 8 adult/yearling moose and 6 Dall's sheep kills each year. Often we have to dispatch a badly mauled caribou or moose during the season. They take refuge in the rivers and lakes, if possible, and usually stand there until they stiffen up or die, and the wolves stay on shore until they do.

I have seen members of a wolf pack turn on one of the members and kill it, for whatever reason. I have also seen evidence where adult members of a pack have turned on the yearlings and killed them during very harsh winters.

Stan Lancaster – Smithers, B.C., Kawdy Outfitters (26 years guiding)

My sons and myself take out wolf hunters in the winter.

Craig Yakiwchuk – Watson Lake, Lone Wolf Outfitters (10 years guiding)

On numerous occasions, we have found 4-8 year old rams killed by wolves. We know this because carcasses are found only by pieces – in other words spread around a large area. Location of kills seem to be in places where sheep do not frequent, but where wolves can ambush them in winter.

The concern with wolves is that they are one of the most efficient killing machines ever devised in nature. They exist and live amongst us, yet we rarely see them. We only see them in the short term (kills, carcasses, remains) or notice in the long term their effects on our game populations. This is the wisest animal on the continent – one that comes as close to reasoning that any animal can. This is what makes it so difficult to hunt, bait, trap or even understand this creature.

Wilf Boardman – Grizzly Basin Outfitters, Cranbrook BC (20 years guiding)

In response to the movie *Never Cry Wolf*, Wilf says "saw the movie and it was quite ridiculous. I've talked to people and they were surprised that it's not true."

Wilf included additional comments by a Mr. Mike Haley who witnessed wolves attacking a grizzly bear with cubs on a hunting trip in 1994 in the East Kootenays. Wilf's son Wade was guiding 2 U.S. hunters when they observed 4 wolves pulling down a big grizzly cub while the rest of the pack kept the mother and other cub busy. "My son, not realizing he could shoot wolves at the time, fired to scare the wolves. They let go of the cub and of course took off. The sow and cubs got away but apparently she was pretty beat, but the way the grizzlies are in that area now and seeing as we can't hunt them, I'm wishing now we hadn't interfered."

Larry and Ingrid Erickson – Alpine Outfitters, Manson Creek, BC (38 years guiding)

I don't believe the present game management policy can work with high or even moderate wolf numbers. As wolves become moderate to high in numbers, what I see is very few calves becoming adults. I see older or unlucky adults falling prey to wolves and I see game numbers dropping to very low numbers. I then see wolf numbers also dropping. As game numbers slowly come back so do the wolves, and game numbers drop again. I also see game, especially moose, very old and using poor feeding areas and showing a general change in behaviour to avoid predation.

Wolves alone can and will keep game numbers low. We have no natural control on wolves on this side of the Rockies (no rabies). You can be sure that game numbers will suffer badly before starvation lowers wolf numbers.

Don Wolfenden – Golden, BC. (30 years guiding)

Wolf predation has reduced elk populations by $\pm 95\%$ in my area, and adjoining Yoho and Kootenay National Parks. No wolves were in this area until 1985. My family has lived in this area since the turn of the century and there never were wolves before 1985.

Stuart Maitland – 100 Mile House (22 years guiding)
One year wolves killed 35 deer in one night on Horsefly Lake when it froze over. It was so easy; they went into a killing frenzy and only ate the noses and lips of some of them. They were not selective, just wasteful.

Roger Williams – Anahim Lake, BC (20 years guiding)
> Wolves are killing too many moose, caribou and domestic cattle. Hopefully, the government will continue the wolf poison program in BC. I feel that this control is very beneficial to all ungulate wildlife and also domestic livestock, outfitters, rangers, and the economy in general.

Alan and Mary Young – Fort St. John (23 years guiding)
Many times have found trophy class animals killed by wolves – 50 inch moose or better. Perfectly healthy animals that became hamstrung, later to die. Wolves do not eat frozen meat so move on to kill again. I have witnessed wolves killing caribou. They herd them into groups and individuals, surrounding them to attack and running them to exhaustion. Wolves kill trophy rams, 7 – 8 years old that are worn down from breeding in November. Wolves kill lambs in the spring. Saw 4 killed last spring in 3 days.

If the people of BC do not allow wolves and grizzlies to be killed, there will be no wildlife left as they have no predators. In the last 4 to 5 years, it has become very evident that the game stocks of all species are being devastated. It is a sad state when wildlife is being managed on political hysteria and government will not stand up to the pressure. Guide outfitters and biologists should be the only people to manage wildlife. Good wildlife management is a balance and the pendulum has by far swung the wrong way.

In 1997 I was witness to an apparent attack by two large wolves as my sheep hunter packed out a sheep head on his back along with meat. The hunter was sitting on a ridge overlooking a basin at about 7000 ft elevation with his back towards the wolves. I was above on a snow pack 200 yards away. The two wolves, a grey and a tan came trotting out of the timber about 500 feet below. They stopped and observed the hunter and then split up running towards the hunter in a half circle formation. They stalked the hunter as if to kill. I killed the grey wolf with my .270, 40 to 50 feet from the hunter. The other wolf turned and ran. The hunter had no clue what was going on.

R. Bruce Campbell – New Hazelton, BC (20+ years guiding)

Fourteen wolves in a pack would have attacked me one cold January night had I not had my candle in a can. Have been attacked by a wolf – it was not tested for rabies. At my trapping cabin on Bell Irving River wolves killed my 3 malamute pack dogs and partially ate them.

Reg Collingwood – Smithers, BC. (26+ years guiding)

My father Dennis Collingwood and his hunting partner were rushed and forced into frigid glacial waters on a beach at the south end of Notase Lake. They claim there were 12 wolves but 9 surrounded them and were aggressive.

Including myself and the guides who have worked for me trophy class animals killed by wolves number: 5 bull moose, 5 Stone rams, 2 goats, 1 black bear. Numerous moose and caribou and few deer. Lots of cow moose.

Ron Fleming – Hazelton, BC. (29 years guiding)

There has been a trend of increase and decrease in the wolf population in the last 30 years that I have been in the north. But every time there is a tip of the teeter-totter, the ungulates (moose and caribou) recover but at a far lower number. An example of this, in my area in the 1930's there was an estimate of 20 to 30 thousand caribou. Their trails are still visible today. Then they went down to nothing and came back in the 1960's, but only to 10 thousand, then down again and back now at 5000. I hate to see them go down again but they are and the Fish and Wildlife Branch is doing nothing. My area is the head of the Finlay River in Tatlatui Park.

Barry Tompkins – Fort St. John, BC (27 years guiding)

Have found remains of dead wolves that have been killed by other wolves but not sure of circumstances. The wolves that the Fort St. John Ministry of Environment have been studying are doing some interesting things: multiple females breeding, intermixing packs and long distance moving.

Bradley Bowden – Quesnel, BC (35 years guiding)

In my opinion, game must be managed – all species, wolves, bears, etc. If we want to enjoy hunting, predators must be controlled. In the timbered country around my area, wolves are hard to shoot, hard to trap and are increasing. In the 1950's and 60's a lot of strychnine and 1080 was used. They should be used again. Local people did it and said nothing. Now with the increasing logging and numbers of people in the backwoods, no control is being done. If something isn't done to control the wolves our hunting privileges will decrease. In the caribou herd in the Cariboo Mountains, wolves are eating the calf crop. Population is increasing.

Bradley Bowden told the author he failed English in grade 9 because he didn't believe what Farley Mowat said in *Never Cry Wolf*. He refers to being "taught" the book.

Dale Drinkall – Mile 422, Alaska Hwy (17 years guiding)

With the wolf population increasing, the ungulate populations are decreasing. I have seen a severe site-specific decrease in the moose population. I have also noticed a site specific decline in the Stone Sheep population mostly on the ewe-lamb ranges where the wolves have been catching the lambs on their summer range, just after lambing season.

It seems that different snow condition in the winter dictates what species are easiest to kill. It is a real misconception that wolves kill only the young or weak. We have found healthy mature animals of all species killed throughout the year.

I have started my own wolf snaring program on my trapline within my guide area since ending our wolf hunts in Region 7 in the northeast. We have successfully taken 57 wolves in 4 years and have learned a lot about wolves during this program. We have noticed that when we have snared a wolf, the pack will keep coming back all winter looking for that wolf. We have noticed bad scars and cuts on their faces either from hunting or fighting when two packs come together.

It is my belief that we have to control wolf populations to prevent them decreasing ungulate populations to a point where it takes a long period of time to come back. I also think that it is healthy for the wolf population if there is some control making hunting the wolf population better. I also believe that more than just one female will have litters each year and the severe increase in wolf numbers would indicate that.

Bernard McKay – Prince George, BC (21 years guiding)

Wolves take as many moose as legal resident hunters. In 1996, two native Indians from Takla Lake (friends of mine), a grandfather and grandson ages 80 and 30 were shooting squirrels with a .22 rifle in the heavily timbered country west of Takla Lake. They were attacked by a pack of approximately 15 wolves. They hid under the canopy of a big spruce tree. The older man shot one wolf between the eyes when the animal came in for them. The men's names are Herb West (33) and Enoch West (80+).

Guy Anttila – Atlin, BC (22 years guiding)

The number of big game killed by wolves by far outnumbers the number killed by hunters. Predators of big game species always select the smaller animals in the group. They therefore predominately end up killing females. We have a lot of habitat for the ungulates which is being utilized only in fractions of its potential. Should we not have the massive uncontrolled wolf predation, I believe it's reasonable to think that the ungulate species could increase 20 fold. On numerous occasions after the moose rut when some bulls are in a weakened state we have found large bulls – 50 to 60 inch spread which were killed by wolves. We estimated the age at 10 to 12 years.

I have found remains of an eaten wolf on 3 different occasions. By the tracks it was obvious that wolves had killed them. These animals had not been shot or wounded as it took place 60 miles from the nearest road and there were no trappers in the area.

Myles Bradford – Dease Lake, BC (35 years guiding)

Wolf population decreased for about 5 years ('85 to 90) as we had partial predator control by John Elliott. Decreasing ungulate populations, increasing wolf numbers. Have found remains of trophy class animals killed by wolves. All ages – mostly in the winter.

9

The Wolf Management Imperative:

In recent years, there have been some excellent wolf-ungulate management studies undertaken in Alaska, Yukon, British Columbia and elsewhere in Canada. These are scientific studies with well documented accurate data.

From Quebec, F. Messier, 1994 (a research biologist) has the following conclusion: "If people were to share the moose with natural predators, and if game managers wanted to avoid major prey declines, the hunter's share of a moose population should not exceed 6 or 7 percent. That is the surplus that could safely be harvested." Obviously, game managers must plan their moose harvests in excess of 7 percent which means that there is no alternative but to regulate wolf and grizzly bear predation.

The wolf biologist, Ludwig Carbyn, 1983 states "From a scientific perspective, wolf control to increase ungulates is perfectly acceptable but is has become socially unacceptable." Socially unacceptable isn't the same thing as right. It is often used by people to give weight to a trend they want to espouse or resist. It is used unintentionally or deliberately to push one's sensitivities on others.

Ralph Archibald's comments at the 1991 North American Wildlife and Natural Resource Conference were as follows: "Wolf control programs have been demonstrated to increase ungulate numbers but they have created such heated public controversy that there is little political will to support them. The seemingly obvious solution would be to abandon these programs, but consider the cost. Our experience shows that if wildlife managers do not address the concerns of those affected by wolf predation, they may take matters into their own hands, at unnecessary cost to wolves and other carnivores." This is sometimes expressed as "shoot, shovel and shut-up."

I suspect that when the provincial government disallowed the renewal of the 1080 control program in 1978 due to public opposition, the covert use of inhumane strychnine poison replaced it. Animal rights proponents should consider the consequences of their intervention in scientifically recommended management methods. Unofficial, covert methods of predator control can make it difficult or impossible to obtain the data needed to scientifically plan succeeding programs when prior information is suspect or unrecorded.

Biologists Robert Rausch and Robert Hinman of the Alaska Department of Fish and Game, 1997 stated that prior to 1959 in Alaska, poison, unlimited wolf harvests, continuous open seasons, summer trapping, den hunting, bounties and aerial hunting did not severely limit wolf populations except in local exceptions. In no instance were the efforts successful in exterminating wolves in any area.

We should never cry for wolves any more than the legal hunter should feel guilty for killing a mallard duck, a coyote or a grizzly bear. Anti-hunting is a contradiction of our predatory evolution and our ascent to top of the food chain. But, we must compassionately and respectfully regard the prey that helped secure our dominant place in the animal kingdom. Animal rights is mythology.

Mech and Boitani, authors and editors of the voluminous book *Wolves*, published in 2003 by the University of Chicago Press, point out that to demand that wolf populations be allowed to continue to increase is not only a false conservation goal but also counter-productive and bound for short term failure. The authors suggest a new public vision that requires a fundamental shift in the way wolves are perceived by those who consider them a symbol of the conservation battle with special rights among all other species. The authors also state that wolves should be managed as part of the whole context and not singled out as a special species. They point out that scientists advocating conservation must strive continually to separate their feelings from their research and their objective knowledge.

Mech and Boitani further state that crucial to wolf conservation is the realization that the wolf is "no longer an animal symbolic of the wilderness. People must come to accept the necessity for management control of wolf populations if the animal's survival is to be assured."

Mech and Boitani comment on Farley Mowat's book *Never Cry Wolf*. "From the best-selling book and popular movie millions of people gained the impression that wolves eat mice, rarely caribou. Author Farley Mowat later admitted fabricating much of the story (Goddard 1996) originally billed as true, to gain public sympathy for the wolf. Mowat succeeded enormously, and decades later this misconception remains. Despite its depiction of fiction as fact, *Never Cry Wolf* probably played a greater role than anything else in creating an icon of the wolf and opposing management control."

In discussing the Media, the above authors of *Wolves* note that the "news are attracted to controversy and wolf recovery, depredations, control programs, and almost any other wolf-related topics seem irresistible."

"The Yellowstone wolf reintroduction was intensively covered by sixty international media. Popular information about wolves is often biased or inaccurate. When wolf stories appear, the extreme views of opponents and supporters of wolves are often highlighted, further polarizing the issue. The way the media covers wolves leaves the impression that they are more of a problem than other animals." Forty non-governmental organizations in North America and at least a dozen in Europe exist to promote wolf conservation (Mech and Boitani, page 298).

Since 1978, newspaper articles opposed to wolf management in B.C. have been extremely counter productive to scientific management of predator and ungulate species. In Alaska where the same situation applies, the Alaska Wildlife Conservation Association prepared an information sheet with headlines as follows:

The Management Imperative and Mythology of Animal Rights

Media Wolf Primer – Let's Get it Right This Time

Much of the unnecessary turmoil and confusion surrounding wolf control programs has been the result of inaccurate and misleading news stories. Following are some examples and basic facts that are as applicable in northern British Columbia as they are in Alaska (from *Media Wolf Primer – Let's Get it Right This Time*).

Authentic Information and Inaccuracies
Courtesy of Alaska Wildlife Conservation Association

Anti-hunters will key on any species to stop management: wolves in Alaska, mountain lions in California, black bears in Colorado.

Without wildlife management there will be no hunting; a goal of animal rights extremists.

Healthy wolf populations depend on healthy moose and caribou populations. No prey, no predators. It's as simple as that.

Hunting, fishing and trapping are environmentally sound activities. They require no pesticides, herbicides or fertilizers.

Sound wildlife management programs, including wolf management programs, will eventually increase wolf populations as moose and caribou numbers increase.

MYTH: *Moose are safe from wolf attacks in deep snow.*
FACT: Moose are more vulnerable to wolf attack in deep, especially crusted, snow.

MYTH: *A control effort taking 300-350 wolves is excessive.*
FACT: Long term wolf harvest of about 1000 per year has had no adverse effect on statewide wolf populations.

MYTH: *Decline of game is due to overhunting.*
FACT: Regulated hunting rarely if ever has caused a decline of moose, caribou, or sheep. Most hunting takes male animals only.

MYTH: *A wolf control program will disrupt pack order.*

FACT: Recent studies show wolf packs are capable of intermingling, breaking up, and continually reforming.

MYTH: *Wolves don't run in large packs.*

FACT: Documented numbers very from one to over 30 animals.

MYTH: *Wolves only kill what they need.*

FACT: Under certain conditions wolves often kill many more prey animals than they can utilize.

MYTH: *Wolf populations are self-regulating.*

FACT: Wolf populations, like other predator populations, will expand to the limit of their food source.

MYTH: *The control programs are an inhumane slaughter of wolves.*

FACT: Shooting wolves is the most selective and humane method of killing wolves.

10

Wolf Predation on British Columbia's Red-Listed Mountain Caribou

The front page of the *Vancouver Sun* newspaper on Sept. 20, 2004 carried the headline, "Fast action needed to save B.C.'s caribou: - forest board." Page 4 included a picture of the Hon. Bill Barisoff, Minister of Land, Water and Air Protection who stated "most of the caribou decline is due to predators." Credible biologists realize this but environmentalists attribute it mainly to logging as occurred when blacktail deer on Vancouver Island were seriously reduced by wolf predation in the 1980's (page 21).

What again appears to be happening is another repeat of organized animal rights and environmentalist activity focusing on forest practices as the cause, rather than predation by wolves and cougars. In retrospect, if the Provincial Government had not capitulated to the antis in 1978 and had renewed the 1080 wolf control program of the 1950's, mountain caribou populations today would not likely be wards of the Federal Government Species at Risk legislation.

It appears that federal and provincial biologists are again expected to fix the problem with politically correct solutions rather than 1080 wolf control. In the writer's opinion, the first course of action should be to reduce wolf numbers that are affecting the mountain caribou by an initial 80 percent and then measure the effect.

There have been successful programs of wolf control in the Yukon, Alaska and in northern B.C. to support this belief. On page 40, Dr. John Elliott describes the spectacular increase in northern caribou on the Horseranch Mountain Range during three years of wolf reduction after which the caribou population plummeted when wolf control stopped.

The excerpts from L.P. Callison's *Wolf Predation in the North Country* in the 30's and 40's include the southern mountain caribou populations that are now red-listed. It is doubtful if logging in the 40's could have caused the caribou to decline at that time. But now environmental groups are attributing the current decline, necessitating red-listing, to forest practices.

Clifford Eagle at Lac la Hache reported lots of caribou and very few wolves in 1930. By 1948 there were few caribou and many wolves in what was apparently still productive forest habitat.

Ted Helset, in South Wells Gray Park, stated that in the 1930's caribou were plentiful with herds of from 20 to 40, at which time he seldom saw sign of wolves. By 1948, he stated that caribou were almost extinct and wolves had increased to the extent they were running in packs of as many as 20. It would appear that the caribou declined even in the absence of logging and human intrusion.

James P. Cochran at Barkerville noted that wolves began appearing in abundance in 1932. By 1948 he estimated the caribou had declined by 85 percent due to wolf predation.

H. M. McNeill, in Central Caribou (Fawn Post Office) stated that there were many caribou on the high mountains and no wolves. By 1948 there were many wolves present in summer when he had been there but the caribou were gone.

Environmentalists should endeavour to understand that wolves, which have a high reproductive capability must be controlled if the less prolific ungulate species, such as caribou are to escape what is referred to as the "predator pit." In addition, there may be limits to enhancing the forest carrying capacity for all southern caribou populations because important social and economic constraints must be considered.

11

Evolution, Genetics, Religion, and Culture

Evolution, genetics, religion and culture may seem unrelated subjects to discuss in a book about wolf management. Let me explain by first saying that I spent the first fifteen years of my life on the Cowichan River, a mile removed from the settlement of Lake Cowichan. Dad was caretaker of the five acre Dunsmuir property south of the village. We lived in a comfortable home surrounded by open grassy fields with the river on one side and the old road from Duncan on the other. Clumps of fir, cedar and maple trees were dispersed over the property but for the most part it was open, resembling the habitat of Palaeolithic ancestors on the open African savannah where visibility was protection from predators and trees were their escape. My early childhood happiness may in part be related to genetic inheritance from whence our human origins began to take shape.

My earliest recollection as an infant was spending time in a temporary play pen against the outside wall on the kitchen side of the house where mother could keep an eye on me. One day she heard me uttering strange sounds not heard before. She ran out and saw a large harmless garter snake curled up in my small pen. I have since learned that one such experience at an early age could be responsible for my genetic induced dislike of snakes originating millions of years ago when poisonous snakes were something to be afraid of by ground inhabiting hominids. As long as I can remember I have had a mild aversion to snakes, as do many people.

Other features of mankind's ancient environment from which we may have acquired phobias are spiders, heights, closed spaces, running water and wolves. The wolf is not one of my phobias but this animal is thought to have been phobic inducing in our ground inhabiting hominid ancestors, whose genes we have inherited.

My mother was raised as a Baptist but as I remember she never attended church after coming to Canada from Sevenoaks Weald in England. She possessed a bible but I never recall seeing her reading from it. Father, a World War I veteran, was not from a religious family either and never attended church to my knowledge. They were not good examples of 20th century religious mythology in which I grew up.

When I was old enough to ride a bicycle to school in the village I acquired friends, some of whom attended Sunday School. Concerned that I might be missing something, but without urging from parents, I decided to find out what religion (mythology) was about and so I attended Sunday School. When I learned that I had only two "choices," either heaven or hell, I decided that religious mythology, as now recognized, was not for me.

About this time, a Mr. Maynard came to retire at Lake Cowichan. He had been an African missionary involved in converting the natives from their belief in primitive mythical gods to that of Jesus Christ, a late comer among most religions. Mr. Maynard brought with him his collection of large hand coloured lantern slides which he had used to convert African natives. I still visualize the scary pictures of humans bent over with the weight and pain from huge bags of sins on their backs which fell to the ground when they accepted Jesus Christ as their new god. The mythical slide shows soon ended due to lack of attendance.

Between the ages of six and twelve, when not in school, I spent many hours fishing in the Cowichan River or hunting with my BB gun for small prey such as winter wrens, chickadees and other small birds. On the one occasion when I shot a red squirrel, I felt even more like a hunter. Little did I know that I was motivated by the genes of primitive hunting ancestors who lived millions of years ago.

A few years later the United Church minister and local boy scout leader was walking past our home on the river and saw me carrying my first .22 single shot rifle. He said nothing but conveyed a clear message of disapproval of hunting by shaking his head as he looked my way. I was a victim of what is now animal rights mythology. Anti-hunters and animal rights advocates are critics of the natural, inherited act of hunting which also includes fishing and trapping.

Not until many years later have I learned why I was apparently such a happy kid with gun or fishing rod, roaming semi-open land similar to that of ancient ancestors and why as an adult, hunting, fishing and nature are spiritual passions with religious connotations that are natural rather than mythical. The simple answer of course, is natural selection, the result of millions of years of natural trial and error as we evolved.

In today's world we have both natural and unnatural environments. All our earliest ancestors evolved in original natural environments over millions of years as small groups of hunter-gatherers. Nature limited their numbers to the carrying capacity of their habitat, no different to that of wild terrestrial species today, including wolves. But modern humans have become less subject to the laws of nature than other species on the planet. Our ancestral hunter-gatherers were completely dependent on food availability as well as naturally existing shelter (caves) to protect them from the adversity of seasonal weather in the raw natural environments.

Anthropologists have traced our evolution by comparing the increase in the size of our skull with that of more primitive *Hominidae*. We may never know what caused our forerunners such as Neanderthal man and *Homo erectus* to become extinct. Theories range from climate change and global catastrophies even to possible conflicts with *Homo sapiens* over millions of years. It is almost impossible to comprehend the length of time it took us to become modern man. Our inability to think in such abstract terms as millennia, natural selection and evolution has contributed to culturally mythical beliefs such as special creation and resurrection. It is important that we understand and accept scientific evidence rather than cultural mythology for our presence on

Earth and our relationship with other species, which mythical animal rightists believe to have rights equal to man.

Species evolution, survival and extinction are natural events as science can prove. Nature recognizes no animal rights or guarantees that emerging species will survive and not become extinct as occurred with the dinosaurs and many other species. We cannot be sure that the same fate will not eventually befall the human animal. Many philosophers have concerns about our long term survival because of our inability to control our population growth for reasons that include mythology and religion.

Nature lost control of our future after the Agricultural Revolution changed us forever. No longer does nature have us living in small groups of wandering hunter-gatherers in balance with the natural environment. Our species has become a victim of culture and civilization.

The next departure from our original natural lifestyle was brought about much later by the Industrial Revolution which has made it possible for mankind to make just about everything that nature never intended us to have, including automobiles, nuclear bombs and ballistic missiles.

Nature is best described as everything not made by man. Only until comparatively recently in our evolution almost everything on the planet was natural but now, natural things such as big trees, pure water and clean air are shrinking. Most humans now live in the unnatural urban environment and express thoughtless opinions of man, the hunter who kills wild animals even though no longer dependent upon them for food but is still driven by genetic inheritance to be a predator and hunt, an inherent act, that became deeply entrenched during our evolution.

Neither the Agricultural nor the Industrial Revolutions made us the most successful species on earth. We had reached this evolutionary level of development many thousands of years before these "very recent" events in our history.

Separated by unnatural cultural differences humankind developed into diverse societies rather than one truly homogenous global society. The many cultures with religious differences developed apart from nature. Culture is said to be "against the grain of nature" and therefore

The Management Imperative and Mythology of Animal Rights

artificial rather than natural. Modern man projects an unnatural image as a result of culture.

But what has the foregoing to do with the imperative or necessity of wolf management, the core subject of this book? Repeated attempts by our scientists since 1970 to regulate the severity of wolf predation on ungulates in our Province have met with failure. They have failed because of cultural opposition in the unnatural urban environment, mainly in Victoria and Vancouver. Anti-hunting and animal rights philosophies of urban culture have been pitted against science by newspapers seemingly interested in promoting controversy rather than objectively reporting scientific facts of natural processes in the interest of conservation. The Sierra Club and Raincoast Conservation Society, for example, endorse unnatural philosophies of animal rights and anti-hunting.

To be politically correct, most democratic governments want to appeal to the electorate and in British Columbia 80 percent of voters live in the unnatural urban environment. In May 1996, the Western Canada Wilderness Committee attempted to seek an injunction supported by a required 10 percent of voters in each provincial constituency to prevent all hunting of bears in British Columbia. The campaign failed because only four percent of rural constituents supported it. It was, however, endorsed in the densely populated urban environments of Vancouver and Victoria. But even then, the Provincial Government endeavoured to further appease animal rights advocates in the urban electorate by implementing black bear hunting regulations to be more politically correct by justifying bear hunting for food consumption purposes. Hence, all black bear hunters were required to retain all edible portions of their "trophy" black bear whether they wished to eat it or not.[4]

In respect to wolf management, it was illegal to use poison to control wolves in B.C. during the first 45 years of the last century. After 15 years of successful wolf management from 1950 to 1965 when poison 1080 was used, political correctness within the urban environment has prevented its use.

---

[4] See *Politically Incorrect: Life and Times of B.C.'s first game biologist*, page 199

There is no better example of urban opposition than the latest (2003) government refusal to support science by removing wolves to protect one of the rarest of all mammals, the almost extinct Vancouver Island marmot. Political correctness and urban culture are to blame.

12

Nature Worship – A Scientific Religion

Nature is described as all things not made by man. Our early primate ancestors in which our genetic character originated, lived entirely encompassed by nature. This was the environment to which our primeval forbearers adapted and from which our evolution continued for millions of years as we gradually became *Homo sapiens*. This primeval dependence upon and struggle with nature was important because it was the challenge for survival that contributed to natural selection and our successful evolution by natural selection.

The predatory habit (hunting) in our development contributed most to our evolution, particularly in brain size compared to other hominid species that are mainly vegetarians. The affinity of early ancestors to nature and hunting is well illustrated in European caves where animals and stylized hunters are depicted on the walls. Randall Eaton suggests that this represents the earliest display of trophyism but it may also be an expression of nature worship.

Contemporary religions and culture have had only superficial influence on human development compared to millions of years of genetic inheritance. Traditional religions, such as Christianity, have experienced problems because they are based on mythology, the human failing "to believe in things that are not so" such as life after death and resurrection.

The renowned religious writer and researcher, Karen Armstrong, in *History of God*, quotes Sigmund Freud; "religion belonged to the infancy of the human race…a necessary stage in the transition from childhood to maturity. He regarded belief in God as "an illusion that mature men and women should lay aside." He also predicted that religious belief represented an immaturity that science would help overcome.

Kenneth Clark in *Civilization*, describes the beginning of a scientific religion, the Worship of Nature. "For almost a thousand years the chief creative force in western civilization was Christianity. Then, in the early 18th century, it suddenly declined and in intellectual society practically disappeared." People, Clark said, are unable to get along without a belief in something outside themselves. "During the next hundred years they concocted a new belief which, however irrational, it may seem to us, has added a good deal to our civilization: a belief in the divinity of nature, the evidence of divine power which took the place of Christianity were manifestations of what we call nature, all parts of the visible world which were not created by man and can be perceived through the senses."

The worship of nature includes many spiritual interests and activities associated with our natural heritage. I believe that my childhood happiness on the semi-open Dunsmuir estate bordering the Cowichan River was spiritually related to genes passed down from ancestors who evolved in the once totally natural environment millions of years ago. I believe too, that genetic inheritance is not only responsible for our obvious attraction to wilderness but is also responsible for our spiritual devotion to hunting and fishing, as well as trapping. At a B.C. Trappers Convention, Douglas Chambers from Kamloops expressed a spiritual interest in trapping. "Most people trap to live but I live to trap," he said.

Many urban inhabitants regularly visit and enjoy natural landscapes and spend days unconsciously "feeling" a spiritual relationship to nature. Unknowingly, they are experiencing their ancient roots to their primeval past when only boundless nature was everything and everywhere.

If Freud were alive today he would probably tell us, that not only conventional religion, but also mythical anti-hunting and animal rights beliefs represent an illusion that scientific thinking will discard. Such unnatural or unscientific beliefs are not part of an intellectual society as they contradict our proven genetic inheritance. Moreover, they oppose the pleasure and consumptive benefits that nature bestowed upon our ancestors and on ourselves, adding immeasurably to protection and appreciation of the natural environment.

Bill Maher, a popular critic of contemporary religion, appearing on the CNN Larry King Live television show used harsh words when he described contemporary religious beliefs as a "neurological disorder, stupid and childish." He referred to organized religion, in particular as a "terrible thing." Albert Einstein noted that only two things are infinite, the universe and human stupidity.

The renowned socio-biologist, Edward O. Wilson, twice Pulitzer Prize winning author of *Naturalist* had this to say about traditional religion. "The myths that dominate our religious beliefs in everlasting life have not helped solve the greatest problem facing mankind – the very survival of our planet and life upon it. He used the term "biophilia" to describe the inborn affinity human beings have for other forms of life. "The most important implication of an innate biophilia is the foundation it lays for an enduring (global) conservation ethic." Starting with educated culture, biophilia is a substitute for religious mythology as we know it today.

13
Mythology of Animal Rights

Prior to the 1950's our provincial game managers experienced little or no opposition from anti-hunters. The concept of animals having rights equal to mankind was never suggested. Reasons for these new beliefs must be addressed if we are to keep the antis off the backs of our game managers. Mythical animal rights beliefs in recent years have made several serious intrusions into game management practices and we can expect more in the future.

(1) Preventing the renewed use of 1080 in northern BC to control wolf predation in 1978.

(2) Preventing the use of 1080 control of wolves on Vancouver Island in 1978. This resulted in the necessary closure of the antlerless deer seasons and reduction in the bag limit from 3 to 2 antlered deer.

(3) Stopping the proposed provision of special wolf traps as an alternative to using 1080 on Vancouver Island.

(4) Pressuring the Provincial Government to stop the Muskwa wolf research using aerial shooting.

(5) Attempt to close all hunting of black bear in BC.

(6) Successfully requiring that all edible meat be retrieved from trophy black bears.

(7) Forcing closure of spring grizzly bear hunting for one year.

(8) Preventing wolf control to protect endangered Vancouver Island marmots necessitating expensive non-lethal attempts to prevent wolf and cougar predation.

Methods employed by animal rights activists have included letters to newspapers, the Premier, MLA's and Cabinet ministers as well as to supportive organizations in other provinces and the American states.

Emotional beliefs that oppose management of wolf predation involve the same arguments against hunting, fishing and trapping. The critics either do not want to know or understand that our evolution has been intimately associated with hunting or human predation. We evolved as hunters and are genetically programmed to be hunters. Moreover, we are the dominant predator in nature's food chain. Modern urban culture has no intelligent basis to oppose the legal consumptive use of wildlife.

Polarization by the media is seemingly encouraged to promote the sale of newspapers. City people often believe that predators and prey animals can co-exist in abundance without intervention by mankind. "Why cannot wolves and moose just be friends rather than killing each other" was expressed by one urban resident.

La Barre (*The Human Animal*) remarks on the characteristics of cultural man being unique among animals in respect to our "practiced ability to know things that are not so." For example, Animal Rightists believe that it is wrong to kill animals even for food, and do not admit to man being a natural predator. In other words, they believe in something that is not so; a belief in animal rights, no-hunting, no-fishing and no-trapping. People for the Ethical Treatment of Animals (P.E.T.A.) believe fishing to be "as cruel as beating a puppy!" Fishing and trapping are simply variations of hunting. Our Palaeolithic ancestors undoubtedly preyed on many mammals and birds for food and clothing and most probably this included fish as well.

Hunting is a natural act for mankind. For millions of years our predatory hominid ancestors, whose genes we carry, were hunters in the fullest sense of the word. In the last 60 years which is less than a second in evolutionary time, all of a sudden, we are expected to

Animal rights glass-covered framed poster placed in Victoria bus stop shelters in Jan. 2005. An emotional seeking exaggeration of the truth (Jenna Hatter, photo).

The Management Imperative and Mythology of Animal Rights

believe that it is wrong to hunt even for food. Quite ridiculous. What other plausible explanation could be responsible for this, other than a serious mythical belief that man the predator is not part of nature. We are not outsiders looking at nature that surrounds us but very much an integral part of it. To some people, especially in the cities, nature is something to look at and enjoy only visually or non-consumptively, while they eat at McDonalds. Their term "socially acceptable" excludes hunting, fishing and trapping. They ignore the fact that consumptive users are also naturalists and by far the foremost conservationists to support sustainability and conservation of the species we hunt as well as protecting a multitude of living things we too enjoy non-consumptively.

Dr. Randall Eaton, a well known American conservationist and educator is a strong advocate of the hunting tradition as an important natural heritage of humankind. He makes the point that our genetic inheritance as hunters has been paramount in our evolution even to the extent of suggesting that people who have never experienced hunting will not know what it is like to be fully human.

In recent years animal activists, primarily in urban society have developed something akin to a profitable industry to sponsor and fund campaigns to attack hunters, fishers and trappers. They employ mythology, misinformation and emotion to solicit financial support for their campaigns. People for the Ethical Treatment of Animals (P.E.T.A.) have a script on their website attacking fishing. "Imagine you're a small child reaching for a shiny new toy. Suddenly, a big hook digs into your tiny hand and snatches you out of the playground and into an environment where you can't breath. This is what people do to fish every day."

Obviously, there is a tendency for many people to think of present day society as new and advanced to the extent that hunting is archaic and no longer part of an appropriate lifestyle. This "belief in something that is not so" has been culturally induced, not genetically inherited. Animal activists believe that hunting, fishing and trapping that result in something being killed, are no longer acceptable pursuits of mankind. This mythical outlook also portrays the wolf as an icon, something special and quiet apart from all other species, a belief that is not so.

Wolves and People

In 1986, biologists Douglas Janz and Ian Hatter reported that since the mid-1980's, deer in many areas on Vancouver Island had declined seriously as a result of intensive wolf predation. "Alternative causes, such as hunting, black bear and cougar predation, habitat and weather, could not account for the deer declines." The polarization of public attitudes to wolf control has been a major factor in delaying reactive management to this problem. They stated that initial, intensive wolf reductions of 80% were necessary to allow a suitable recovery period (5-10 years) for most deer herds on northern Vancouver Island.

The biologists pointed out that shooting, trapping and especially humane poisoning had been used to control wolves elsewhere in British Columbia. Experience on Vancouver Island, however, indicated that regulated hunting and trapping were not generally effective. Aerial shooting is not feasible on Vancouver Island.

The use of sodium monofluroacetate (1080) was considered the most effective control method due to its selectivity for canid animals. Using small pieces (two to four ounces) of meat bait treated with one lethal wolf dose (13 mg 1080), the baits are completely safe to humans and other wildlife species similar in size to wolves. Burying baits reduces the chance of removal by small mammals and birds. Due to the dilution factor, secondary poisoning (feeding on a poisoned carcass) is virtually impossible. A hearing by the Environmental Appeal Board in 1980 on the use of 1080 for control of problem wolves and coyotes concluded that the use of poison "as outlined by permit, posed little or no threat to the environment and was the most effective and economic method of predator control."

In addition to environmental concerns, opposition to the use of 1080 is often based on the claim that it is inhumane. Because 1080 is relatively slow in taking effect (two to three hours), however, does not imply it is less humane than faster-acting poisons or methods. As 1080 acts to inhibit glucose metabolism, there is a latent period of about two hours after ingestion during which symptoms are absent. This is followed by a 10 – 15 minute period of over-reaction (convulsions, etc.) prior to coma and death. Although pain in animals cannot be measured, recent research indicates the use of 1080 is a tranquil process in which the victim is oblivious to the external signs that distress human observers. Human victims of accidents or suicide attempts for example, reported no memory of convulsions nor did they feel any pain (F. Tompa, pers. comm.). Dr. Tompa was a Fish and Wildlife Biologist in Victoria, involved with problem wildlife management.

The results of scientific studies on Vancouver Island referred to above and the recommendations of the Environmental Appeal Board were not acted upon by the Provincial Government. The use of 1080 was not approved by the government of the day.

Knowing that some successful local control of wolf numbers in the Nimpkish area had occurred by trapping during wolf-deer studies, the Fish and Wildlife Branch, with no other option available, recommended that trapping efforts be expanded in attempts to reduce wolf numbers on Vancouver Island. Apparently, Government approval was obtained to purchase a substantial number of Braun wolf traps for distribution among trappers on Vancouver Island. The writer was supplied with 15 wolf traps to use on his registered trapline in the Jordan River area on southern Vancouver Island. Two wolves were trapped but the following winter the traps were recalled by the regional office in Nanaimo for use by trappers in areas where there were more wolves. Apparently, use of the traps was not restricted to the regular winter season when wolf pelts are prime and commercially valuable to trappers.

It appears that the government again capitulated to animal rights objection, this time to extended wolf trapping and to government having supplied traps to trappers to kill wolves to benefit deer hunters.

The government's "politically correct" rejection of the Fish and Wildlife Branch's request to use 1080 to prevent excessive wolf predation on Vancouver Island in 1978 resulted in the deer harvest plunging from 14,000 deer harvested in 1980 to only 6,000 in 1988 causing the decline in deer hunter expenditure from almost 7 million dollars in 1981 to less than 2.5 million in 1988. This was an economic and social cost of 4.5 million dollars for Vancouver Island (see page 21).

The decline in the deer population made it necessary for the Fish and Wildlife Branch to severely limit the opportunity for hunters to harvest a deer by closing the seasons on all antlerless deer and reducing the annual bag limit from 3 antlered deer to 2. In other words, there were not enough deer to meet the demands of both hunters and wolves without further reducing the deer population.

It is important for the reader to understand that animal rights opposition to hunting, fishing and trapping is often an organized or solicited letter writing campaign to newspapers and elected members of government. Objection to renewing the 1080 program in 1978 was organized, as was the aerial wolf control research in the Muskwa (please refer to pages 22 and 35).

It is a common misconception on part of most hunters, fishers and trappers that governments proceed on good scientific information and advice even when opposed by animal rights activists.

After retiring from government service, I was elected to the board of directors of the BC Wildlife Federation in 1982. Subsequently, on behalf of the federation, I wrote to the newly appointed Minster of Environment, the Honourable Stephen Rogers, concerning his apparent inaction in not supporting the Fish and Wildlife Branch request to use 1080 to control wolf predation on Vancouver Island and re-instate the northern 1080 program of the 1950's when wolf populations rebounded in the 1970's.

Rogers reply to my letter was politically correct but not convincing that the Provincial Government was serious about the social and economic need to control wolf populations. "I feel", said Rogers, "that you have not quite caught the measure of contemporary opposition to any degree of wolf control, and particularly to the use of poison. It is certainly true that much of the "hands off" outcry stems from what I have termed an unrealistic Walt Disney attitude, and we all know its sources. This apart, there is a considerable degree of expertise and intelligence also lined up with the many people who quite sincerely wish us to leave the wolves alone. While we cannot go all the way with these people, setting wildlife management policy in today's climate of opinion calls for careful balancing and, often, considerable caution."

Stephen Rogers should have known that the letter writing campaigns of the late 1970's and early 80's were solicited and organized by animal rights activists and not the result of voluntary individual belief. In future, threats from organized activist groups should be evaluated objectively by senior ministry advisers for their face value.

The Management Imperative and Mythology of Animal Rights

In December, 1979 the former Minister of Agriculture, Cyril Shelford, wrote to the Honourable Stephen Rogers as follows:

> No doubt you have heard me criticize Game Management in this Province for the last ten years, since Predator Control and Game Management were discontinued. I remember what the Swedish Game Managers said to me in 1953. There is no such thing as Game Management without Predator Control and, with the tremendous increase in game numbers since that time, they have proven their points as ours have continued to decline since Predator Control was discontinued. Up to that time, our management in the Province was second to none in most areas. We have always been fortunate in having very dedicated people in the branch – even though I haven't always agreed with them.
>
> It is frustrating for many people like myself who were guides and trappers for many years and have seen the game population nearly destroyed in the late 1930's and 1940's – then saw the dedicated people in the Game Branch get out and resolve the problem with Predator Control. We saw deer, caribou, sheep and goat come back in large numbers even with the increase in hunters; and now we have to sit back and hear sincere, but poorly informed, groups like Green Peace, the Telkwa Foundation, Sierra Club, most of the media and thousands of citizens, mainly from the city, who simply don't understand Game Management or the normal change in wildlife populations.
>
> It is known that game populations build up until the predator numbers increase to such a point that they wipe out most of the game – then they also die for lack of food and the next cycle begins. These groups don't know what it's all about and while our game is being destroyed they sit back and say either we don't know why the game population is going down, or blame it on the hunters. They completely disregard past experience and common sense in the field of Game Management. Common sense is very rare today due

to a faulty educational system on this question.

At a time when more and more people live in urban areas with no opportunity to observe game in its natural state, few teachers, through no fault of their own, in both grade schools and universities, have any background in the subject and rely entirely on books and TV programs of questionable value. The Grizzly Adams – Farley Mowat type stories lead thousands of our students down the wrong path and leaves them sincerely believing in a myth.

You still have many good people in the Fish and Wildlife Branch who know all about the same problem we faced in the 1940's and early 1950's, also, how they resolved it. I am certain they, like myself, feel terrible about the loss of our game and having to stand by doing nothing to save our beautiful animals.

Jack Kempf gave some good advice to the Hon Pat McGeer some time ago regarding a review on what we are teaching in our schools and universities on wildlife and game management because, in my opinion, the education in this field is mainly of a minus value. It would be far better to have nothing at all than what is being taught today.

I trust you will review this whole question as quickly as possible to see if the little game we have left can be saved - even though I realize that with the myths taught on this subject in our educational system and the lack of knowledge by the general public it will be difficult to explain a management program of control not elimination.

Kind personal regards and best wishes in your new job.

Cyril Shelford

14

The Wolf's Role in Evolution, Genetically Induced Phobias and Cultural Dilemmas

Millions of years ago our ape-like tree living hominid ancestors would have looked down from the safety of the trees to see large four legged predators looking up at them. Some were probably wolf-like animals similar to modern wolves. When we became bi-pedal ground inhabitants our only escape before we had weapons, was to climb a tree if attacked by a predator.

As our primeval ancestors evolved from life in the trees to ground inhabiting hominids they were preyed upon and eaten by relatives of present day predators such as wolves and bears. The physically best would naturally have had better chances of survival. The more alert, cautious hominids who stayed within reach of trees also would be more likely to survive predator attacks and pass on their genes to evolving generations.

Apart from possibly contributing to human evolution the wolf, at some stage in our development may have become entrenched in our genetic inheritance and passed down to the present as a phobia. In his book, *Naturalist*, Edward O. Wilson refers to features of mankind's ancient environment from which we have acquired phobias of such things as spiders, heights, closed spaces, snakes and wolves. Long ago evolving hominids were killed by wolves as well as bitten and killed by poisonous snakes. Such phobias are thought to have become entrenched in our genetic inheritance and passed down to the present. Bedtime stories in almost every nation and language have taught children to "beware of the wolf."

In their 448 page, all encompassing book, *Wolves*, Mech and Boitani quote Schaller, Lowther and Peters in an interesting comparison between early humans and wolves which occupied similar ecological niches. "Both were broadly adapted predators of large herbivores and hunted in family groups. People and wolves lived in loosely analogous societies that shared such characteristics as pair bonding, staying together year-round (not just for a breeding season), extended family clans, group cooperation, communal care and training of young by both males and females, group ceremonies, leadership hierarchies and the sharing of food with kin. Like early humans, wolves defend their territory from other packs. Although wolves and humans probably scavenged from each others kills, we do not know whether Palaeolithic people saw wolves as competitors."

When early man was able to exert dominance over the wolf and other large predators he rose to top of the food chain. He developed a remarkable relationship with the wolf. It presumably began when he arrived back at the cave with a wolf pup or two, already weaned and not dependent on the mother's milk. It also has been suggested that he may have captured a pregnant female and kept it captive to feed its young. Another possibility was a lactating female with young. We will never know, but we do know that *Canis domesticus* is the wolf's ancient gift to mankind. It is common knowledge that all our domestic dogs originated from the wolf. But it was not until *Homo sapiens* began to live in settlements and practice agriculture that we first began to develop different breeds of dogs. It seems that the possibilities were almost limitless when we compare a miniature Chihuahua with its wolf ancestors and the many shapes and appearances of family pets and working canines, loved the world over.

The Management Imperative and Mythology of Animal Rights

Our dilemma in understanding the wolf in relation to competition with ourselves, the dominant hunting predator, we have some problems. Not the least of these is the kinship we have with all "domestic wolves", large and small that live with us, eat with us, sleep with us, hunt with us and adore us, but in many ways still behave and may even look like their progenitor *Canis lupus* from which they were genetically "manufactured."

Although the wolf is untamed, large, wild and free, some of us may have a tendency to relate to the "original dog" consciously or unconsciously as we do our own. Although these friendly looking wild canines appear in works of art and adorning our walls, we have to deal with such innocent killers when in need of control; no longer an imagined friend but a competing predator. Fortunately, evolution has allowed us to become the only predator with a sense of compassion for all other species, including other predators.

Some writers and artists have unfortunately eulogized the wolf, to the extent that it has become an icon. They see the wolf only as an animal much less destructive of the ecosystem and biodiversity than ourselves. As a result, there is often more than a hint of protectionism in books on wolves. Writers are awed, naturally, by the ability of a single wolf to kill large prey such as a bull moose, using only endurance and specialized cusped or predatory canine teeth, which we still retain, in rudimentary form.

The wolf, with its greatly superior reproductive ability among its ungulate prey enables it to inflict damage on our domestic animals and mutual prey and therefore often interferes beyond acceptable limits with our stewardship or conservation of prey species. It is the urban animal rights advocate in particular, whose judgement is flawed in mindlessly opposing control of wolf populations that need to be managed in the interest of biodiversity, our abundant natural heritage and the wolf as well.

Those who simply suggest that we let nature take its course regardless, must realize that we hunters, fishers and trappers are important participants in nature and an important force for wildlife stewardship in a conflicting economic world.

15

Understanding the Human Predator: Why We Hunt, Fish and Trap

Randall L. Eaton, an avid hunter himself, says that for the most part, hunters fail to articulate the inner side of hunting and communicate it effectively to anti-hunters, who refer to hunting as cruel, archaic and unnecessary. We are not good at explaining what hunting does for us as human beings and so he says, "we leave out the very heart of hunting." We must accept blame for this but I hasten to suggest that anti-hunters and environmentalists who criticize us, are even more to blame for not trying to understand that hunting is a natural act driven by our genetic inheritance. If it were not for hunting, mankind would not exist! The morality of hunting which includes fishing and trapping is attacked by animal extremists and for the most part is misperceived by the general public.

Ortega Y. Gassett, the author of 'Meditations on Hunting' is the world renowned authority on hunting. He and other philosophers point out that at birth life is given to us empty and we have to fill it ourselves. Lesser animals, on the other hand, are programmed with pre-determined conduct or behaviour and for them life is never empty or undetermined. Man, on the other hand, has lost this system of instincts

The Management Imperative and Mythology of Animal Rights

or retains insufficient to provide a predetermined pattern of behaviour. The dominant activity of our primitive ancestors, more important that all others, was hunting. Ortega says that our life consists of conflicting occupations, the laborious and the pleasing. Hunting is a natural form of happiness which we inherited genetically and which fills the pleasurable side of what began for our species as an empty life. We are genetically programmed to experience hunting as a pleasurable part of life. Ortega scorns recreation or diversion as the reason why we hunt. This description implies comfortable situations, free of hardship, risk and not requiring great physical effort and concentration. As carried on by a serious hunter, it involves all these things including support for a predator management strategy when necessary.

Anti-hunters must realize that the confrontation between man and animal has precise boundaries, beyond which hunting ceases to be hunting. Should the trapper or fisherman (also a hunter) let loose his technical superiority by using poison he ceases to be a hunter. The biologists who control wolf predation by humane aerial shooting find no pleasure as a hunter and usually dislike what they have to do. It is unpleasureable work, that is not hunting. Hunting must be a fair chase situation in which one animal strives to hunt, while the other strives to not be hunted. The superiority of the hunter over the prey can not be absolute or it would not be hunting. Many people believe that the modern hunter has all the advantage. This can not be so or he would not continue to hunt. The hunter, in restraining his predatory ability to capture, deliberately weakens himself in relation to the prey. For pleasure, he returns to nature less capable than he could be as a predator. Ortega says, "this indicates why hunting is such a great delight and pleasure for man." He deliberately becomes closer to nature by abdicating his human potential to capture. Dedicated hunters know what is meant by this natural participation with the animal. It explains why archery is favoured by some hunters. The highest satisfactions arise from renunciation of superiority over the prey. We derive pleasure from being natural and it is most natural for us to hunt. Killing a prey animal is a natural human act.

The men who each day killed dozens of helpless harp seals on the flow ice by clubbing them were not hunters, nor is the man in the abattoir or the person dropping poison baits from an aircraft to control wolf predation, a trapper.

One of the first and most pleasing acts of hunting is to locate the prey animal. If it were not necessary to locate the prey it would not be hunting. The fact that man hunts presupposes a scarcity of game. If game were superabundant, man's hunting behaviour would not exist. Hunting has evolved because of the relative scarcity of game, says Ortega.

Hunting, fishing and trapping are a pleasure to many people and an obstacle to ecologically devastating human developments. Ortega refers to a conscious and almost religious humbling of man, limiting his superiority and lowering him towards the animal. Randall Eaton, also refers to the obvious religious or spiritual connotations of hunting.

The true hunter, says Ortega, takes pleasure in his artificial return to nature; hunting is the only occupation that permits him something like a "vacation from his laborious human condition" which civilization has imposed upon us.

"The hunter is, at one and the same time, a man of today and one of ten thousand years ago. The residual primitive genetic instinct that man has retained is the reason why today, at the end of innumerable millennia, he experiences hunting as a form of happiness."

Criticism from anti-hunters, animal rights advocates and some environmentalists is mostly mythical and unscientific. They fail to respect evolution and man's position as a predator at top of the food chain. Moreover, they fail to realize that man is the only predator with the ability to reason and therefore show compassion for the prey. In the predator-prey relationship in nature there is no cruelty. Hunters, fishers and trappers are natural predators with a reasoned aversion to cruelty, as well as a conservation ethic.

16

Two Accounts of Individuals Treed by Wolf Packs in Northern B.C.

Prospector Fights off Attack
From Times Colonist – July 17, 2004

The gold miner was camping in the northwestern B.C. wilderness when a pack of wolves attacked and killed his dog, Buddy, then stalked and chased him through the woods for days. Luckily, he made it back to the safety of his family in Chilliwack last weekend with his terrifying tale.

In late June, Pinette, 34, had been camping at Eva Lake near Atlin, B.C., 60 kilometres south of the Yukon border. He had hiked into town and decided to return to camp by a different route in late June. This trip would prove to be the most unusual in a lifetime of hiking and camping.

A few days into his hike, Pinette found himself surrounded by a wildfire. "Basically in every direction I looked, I seen smoke," said Pinette, speaking from his sister's home in Chilliwack. He decided to make camp for a few days until the fires died down.

On the last night, Pinette said his dog, Buddy, "went snaky" on him, growling, snapping and rolling his eyes. "It was a struggle to hold onto him." Pinette had adopted the 10-month old Husky-Shepard puppy from the Atlin animal shelter.

"He was so smart...I'd explain my plans verbally to him every morning so he'd know what we're doing," said Pinette, adding that Buddy had turned into a "total bush dog" who would dig his own holes when Pinette went prospecting, follow him along the trails, and curl up with him in a sleeping bag to share body heat during the nights.

But at 3 a.m. that morning at the end of June, Buddy was attacked and killed by a pack of 10 wolves just a few metres from the tent. Pinette managed to scare the wolves away with a can of bear spray, but not in time to save his dog.

"When I turned around, all I could see was two eyes in the bush," he said. For the next two hours, Pinette could hear the alpha male "clacking" his teeth.

A short while later, Pinette saw a wolf's paws digging under the side of his tent. "I started hitting the paws with a hatchet," said Pinette, hoping to drive the animal away. When that didn't work, he cut a hole in the tent and used bear spray to scare them away. "I didn't want to hurt them at first," said Pinette, but eventually he went at them "full force" with the hatchet and bear spray, managing to hit a few.

When Pinette accidentally knocked down his tent, another four or six were trapped inside, he said. "I was hacking and smashing at those too," said Pinette. "I think I managed to kill all of them."

The wolves stalked him as he moved on for safety.

His adrenaline pumping, Pinette took shelter in a tree. "I had my back turned (as I was climbing up the tree) and one came right up behind me and scared me," said Pinette, who dropped his hatchet and fire kit in shock.

A terrified Pinette spent a day and a half up the tree with

no food and water, because he could still see three wolves waiting below.

"I was almost ready to give up," said Pinette. "The first thing I envisioned was my mom's meat pie at Christmas, and I pictured my nieces and nephews and family. Somehow I found the drive to keep going. My feet hurt, my back ached….Every time I'd fall down, it was like something was pushing me and telling me to 'Just get up and go', he said.

The next few days were a blur. He moved toward the sounds of power saws being used by the firefighters, and finally came across a river. "I stood in the O'Donnel (River) for three hours just drinking water," he said.

On the ninth day, Pinette stumbled onto a gravel road and recognized a creek. He then walked for 12 or 13 hours, and slept in an old school bus that night.

The next morning, he came across a group of botanists, who rushed him to the Red Cross outpost in Atlin.

George Ball, Telegraph Creek

In the last hundred years or more there probably have been many unrecorded instances of human-wolf encounters resulting from communication problems or simply because the experiences were accepted as part of life in the wilderness. Georgiana Ball describes her father's unrecorded experience in March 1942.

George Ball encountered a pack of wolves near his ranch close to the Stikine River 12 miles from Telegraph Creek, BC. He was returning home late in the afternoon from blazing out a new hunting trail in his guiding territory. He was on snowshoes, carrying only a short handled Hudson's Bay axe. He had left his rifle hanging on a tree limb one mile ahead down the trail in order to be less encumbered while limbing trees and marking the new route. As he entered a small sparsely treed opening he heard whimpering sounds and saw two wolves which he realized were probably scouts communicating with the rest of the pack. Without his rifle

Wolves and People

he sensed that he could be in trouble. He snowshoed as fast as possible to the nearest pine tree of a decent height, and kicked off his snowshoes just as nine black wolves entered the partial clearing in a semicircular formation. He reached the lower limb of the tree carrying his axe, just in time to climb out of reach. The wolves began milling around the base of the tree looking up at him. He cut off branches and threw them down in hopes of scaring them away. It was cold and George was fearful of falling asleep with hypothermia. He decided that if he had to, he would use his belt to tie one arm to a limb to prevent falling

After about 1 ½ hours, it was dark. The wolves disappeared but he stayed in the tree for another half hour just in case they were still in the area.

With snowshoes on he headed back to his rifle and then another two miles to home. Agnes, his wife, saw that he was noticeably shaken, when she asked why he was so late getting back. He tersely replied, "I was treed by nine black wolves."

This was a wilderness experience with wolves, which in those earlier years, never reached the newspaper. Georgiana, still has vivid recollections of her father's description of the encounter.

17

Conclusions and Comments

It should be obvious that recent Provincial Governments have given too much attention to organized emotional and mythical campaigns of urban animal activists. These have opposed traditional hunting and trapping activities which are so much a part of our rural lifestyle and spiritual enjoyment of nature.

British Columbians are endowed with a rich variety of fish and wildlife resources that equal or exceed those of most other provinces. The social and economic benefits are enormous. Unfortunately, this is not recognized by the public except those in sportsmen's organizations. Anti-hunters and animal right activists target the less informed general public with newspaper articles to achieve support for their mythical viewpoints.

It is important that the general public as well as our MLA's understand the basic scientific management that regulates the consumptive use of wildlife that benefits and enhances the social and economic life of our province. The control of predation is a part of this management. The successes achieved by animal activists in recent and past years in opposing predator control is related in part to the lack of understanding on the part of the people we elect to government. However, we individual hunters, fishers and trappers also share part of the blame along with the elected people we look to for help. We are not good at articulating what hunting and fishing mean to us and how natural and healthy it is for us to be a spiritual part of nature.

Our MLA's need to hear more from us, a great many of us. In the days of Attorney Generals Gordon Wismer and Robert Bonner we were encouraged to talk to our respective MLA's about game management. Our elected members still need to hear from us as individual hunters, fishers and trappers. We are their constituents whether we live in the city or in rural areas. They need to hear how we individually feel about the mythical beliefs of animal rights people who would like us to stop all hunting and trapping. How else can MLA's, especially those newly elected, sort out the conflicting opinions they may receive from anti-hunters and animal activists.

The majority of letters to newspaper editors about hunters are from anti-hunters and extremists. It is exceptional to read a newspaper column originating from a hunter. This is not helpful. The general public needs to become better informed than they are about issues involving game management. The antis have an advantage due to the paucity of public information emanating from government of an educational and scientific nature.

Before the availability of the internet, almost every state wildlife agency in the USA had a quarterly information and education publication similar to British Columbia's *Wildlife Review* which was dropped by former administrations. The writer believes that Information and Education (I and E) are essential to achieving public support for science based policies of our biologists who deal with game management and consumptive use of wildlife. Knowledge engenders public interest and understanding of the age old evolution of mankind as a predator that we translate into hunting, fishing and trapping.

References

1. Campbell, J. 1962. The masks of God: Primitive mythology. Penguin books.
2. Carbyn, L.N. 1983. Management of non-endangered wolf populations in Canada. Acta Zool. Fenn. 174:239-243.
3. Dobshansky, T. 1962. Mankind evolving. Yale University Press.
4. Eaton, R.L. 1998. The sacred hunt. Sacred Press. P.O. Box 490, Ashland, Oregon, 97520.
5. Goddard, J. 1996. A real whopper. Saturday Night III(4) 36-50, 52,54,64.
6. Hatter, I.W. 1984. Effects of wolf predation on recruitment of black-tailed deer on northwestern Vancouver Island. Master's Thesis, University of Idaho, Moscow, Idaho. 156 pp.
7. Hatter, J. 1997. Politically incorrect. Desktop Publishing, Victoria, B.C.
8. Jamsheed, R. 1976. Big game animals of Iran (Persia). Master's Thesis. Colorado State University.
9. Janz, D. and I. Hatter. 1986. A rationale for wolf control in the management of the Vancouver Island predator-ungulate system. Wildlife Bulletin No. B-45. Ministry of Environment. 35 pp.
10. La Barre, W. 1968. The human animal. University of Chicago Press.
11. Lawrence, R.D. 1986. In praise of wolves. Totem Books.
12. Mech, L.D. and L. Boitani. 2003. Wolves, University of Chicago Press.
13. Messier, F. 1994. Ungulate population models with predation; A case study with North American moose. Ecology 75:478-488.
14. Mowat, F. 1963. Never cry wolf. McClelland and Stewart, Toronto.
15. Ortega Y Gasset, J. 1972. Meditations on hunting. Charles Scribners Sons, New York.
16. Rausch, R.A. and R. A. Hinman. 1977. Moose management in Alaska. An exercise in Futility?

17. Wilson, E.O. 1975. Sociobiology. Harvard University Press, Cambridge, MA.
18. Wilson, E.O. 1995. Naturalist. Warner Books, New York, N.Y.
19. Wilson, E.O. 1998. Consilience, the unity of knowledge. Alfred A. Knopf, Inc.
20. Young and Goldman. 1944. The wolves of North America.
21. The emergence of man. Time life books. New York. 6 vols.
    a. Life before man
    b. The first men
    c. The missing link
    d. Cro-magnon man
    e. The Neanderthals
    f. The first farmers

Appendix 1. Wolf Questionnaire to Guide-Outfitters in B.C.

Dear Guide Outfitter:

This questionnaire is to request information for a book I plan to write with a different perspective on wolves. Those who have read my book *'Politically Incorrect'* will know where I stand on the importance of wolf management (control). This has not been in effect in British Columbia on an appropriate scale since the 1940s and '50s. I was BC's first game biologist in 1947 when the last predator control program began and continued for nearly 20 years.

I would very much appreciate your answering the following questions. It is important for those of us who hunt and guide to record factual information from people who derive income from guiding and not simply observations of writers who tend to show little or no concern about hunting and the economic impact of predation on game species.

For example, last year in Yellowstone Park where prey is abundant, three female wolves in one pack raised a litter of pups – so much for the popular belief that only the dominant male and female breed.

1. Name of Outfitter: _____

   Mailing Address: _____

   Telephone Number: _____

   Radio-telephone Number: _____

2. Number of years guiding: _____

Wolves and People

3. Big game species hunted in your area (*in order of abundance*):

   1st: _____ 2nd: _____ 3rd: _____ 4th: _____

4. Big game species hunted in your area (*in order of economic importance*):

   1st: _____ 2nd: _____ 3rd: _____ 4th: _____

5. What has been the trend in wolf numbers in your guide territory since you began guiding? (circle one):

   Increasing   Decreasing   Fluctuating   Unknown   About the same

6. What has been the trend in wolf numbers in your guide territory since 1990? (circle one):

   Increasing   Decreasing   Fluctuating   Unknown   About the same

7. What has been the largest pack of wolves you personally have counted?

   From Aircraft: _____   From the Ground: _____

8. What is the largest pack reported to you by another person or persons?: _____

9. Are you concerned about wolf predation in your guide territory?
   _____

   If so, what effect do you believe wolf predation is having in your guide territory? (*please use back of page, if necessary, for comment on this question*)

10. Do you believe wolf predation has reduced the abundance of fur species such as beaver, lynx, fox and coyote in your area?:

11. How many domestic animals such as horses, cattle, dogs, etc. have you lost to wolves?:

    Horses: _____ Cattle: _____ Dogs: _____ Other: _____

12. Do you know of any person being attacked by wolves? If so, please give as much detail as possible on back of page. Was the wolf killed and tested for rabies? _____

    (There are now documented records of wolves having attacked people)

13. Have you found remains of trophy class animals killed by wolves? If so, please use back of page for details such as estimated age of the animals and time of year.

14. Have you observed round *firm* cysts (hydatid cysts) filled with fluid embedded in the lungs or liver of moose, elk, deer, caribou, mountain goat or mountain sheep?:
    _____

15. Do you know of any person who has been diagnosed as infected with hydatid cysts in their lungs and liver?
    _____

    (The eggs of the adult tapeworm parasite are spread by wolves to ungulate animals and indirectly to domestic dogs which have eaten infected lung and liver tissue and hence to humans. Direct infection also can take place from careless handling of wolf carcasses.)

16. Have you seen or heard of members of a wolf pack in your guiding territory turning on a disabled or wounded (e.g. shot) member of the same pack? _____

Wolves and People

(please give additional details on back of page if answer is yes)

17. Do you possess or know of persons who have photographs of game or domestic animals killed and fed upon by wolves? _____

    If yes, please supply, if possible, names and addresses of persons who may have such photos or know of others who may have pictures that may be copied.

18. Have you talked to anyone who has read Farley Mowat's book, *Never Cry Wolf* and believed it was a true account of the author's experiences?_____
    (Never Cry Wolf was published in 1963).

19. Have you talked to anyone whose children have been instructed by their teacher to read Farley Mowat's book, *Never Cry Wolf?*_ _____

    (It was placed on the non-fiction, optional reading list, for grade 9 students in BC schools. In some instance, teachers made it compulsory reading in grade 9. It was described as the true experiences of the author. It is almost pure fiction and likely has contributed to our legacy of "wolf lovers" in cities as well as an unknown number of misinformed politician and their constituents who oppose any killing of wolves for game management).

20. Do you know anyone locally who has raised a wolf pup to maturity (2 years)? _____
    If yes, pleas provide name and address on back of page if possible.

21. Do you know anyone who owns or has owned a hybrid dog-wolf cross other than a wolf and husky or malamute cross? _____

    If yes, please provide name and address on back of page.

22. I would like your permission to quote you on your observations.

Permission Given: _____ Permission not given:_____

23. If you have any additional comments, whatsoever to make about wolves and wolf predation, please do so. Here's hoping you have a successful season this fall. Many thanks for your help!

Sincerely,

Jim Hatter

Appendix 2. Letter to Pesticides Directorate, Ottawa

21 July 1987

Mr. George Laidlaw
Pesticides Directorate
S.B.I. Building
2323 Riverside Drive
Ottawa, Ontario, K1A 0C6

Dear Sir:

I am a retired former director of the British Columbia Fish and Wildlife Branch. I write this letter out of interest for the seeming inability of our Wildlife Branch to obtain the necessary approval to use 1080 (sodium monofluroacetate) to help control excessive numbers of wolves on Vancouver Island and other locations in British Columbia. Can you answer for me the following questions?

1. Has the Province of British Columbia requested approval for its predator control specialists to use 1080 to control predation on wild ungulates?
2. Is the information I have that the Federal Government does not approve use of 1080 to control wolves to protect ungulates correct? If so, can you advise me why this should be the case with species whose management falls under provincial jurisdiction?
3. Is the earlier injunction by Greenpeace against a B.C. permit renewal to use 1080 for wolf control still in effect and has it been contested?
4. To what extent do you feel public opinion influences the Federal Government in its issuance of provincial pesticide permits to use 1080 for wolf population control?

I would appreciate your candid response to my questions.

Yours truly,
Jim Hatter

Appendix 3. Response from Pesticides Directorate

Dr. J. Hatter
3931 Tudor Avenue
Victoria, B.C.
V8N 4L5

January 8, 1988

Dear Dr. Hatter:

Re: Registration and Use of Sodium Monofluroacetate

Please excuse the delay in responding to your July 21, 1987 letter.

I shall try to address your concerns. The B.C. government has received registration for sodium monofluroacetate as reflected in the enclosed label.

We have been aware that B.C. is considering the use of the active in the wild area with respect to ungulate management.

Wildlife falls within the jurisdiction of the provincial governments.

We would be happy to review a requested amendment to the use pattern. Our concern is limited to the Pest Control Products Act & Regulations where safety, merit and value are the key operative words used in determining registration and use of control products in Canada.

You may wish to request clarification on the Greenpeace injunction against B.C. with your former colleagues in the Wildlife Branch.

In your fourth question to what extent do you feel public opinion influences the Federal government in its issuance of a provincial pesticide permit to use 1080 for wolf population control? I believe that as the evaluation officer responsible for this area, I must limit my input to the science and documentation necessary to fulfill my responsibilities under the Pest Control Products Act.

      Yours sincerely,
      George W.J. Laidlaw
      Vertebrate Pest and Insecticide Evaluation Officer
      Product Management Division

Appendix 4. Rio – The Pet Alaskan Wolf

One evening in the summer of 2001, I had a remarkable experience. I was returning to our cottage in Qualicum Bay after taking my German wire-haired pointer for an evening run along the Big Qualicum River. Approaching me on the trail were four people, a couple with two small children and a huge "dog" on a short leash. As I came closer, I was surprised to see an animal that stood nearly waist height to the adults. I stopped to say hello and said to the man "your dog looks like a wolf." "He is a wolf," the owner replied as I withdrew my hand from the big animal that had just licked my fingers. "Rio is an eight year old Alaskan wolf" replied the owner.

As it was getting dark, I arranged to visit Rio's owners the next day for more information. I learned that they had only recently acquired Rio from another local couple who had kept him chained up for 3 years with little or no exercise other than from a long running wire beside their house. They had also owned a black dog-wolf hybrid that was destroyed because it had attacked and bitten someone.

Rio was taken from a den at the age of 4 weeks, probably in Alaska. Before coming to BC, he was a house pet for 5 years somewhere in the State of Oregon. Apparently, he had been well treated, even sleeping on the owner's bed at night. He had only recently been acquired by the people I met on the trail under the condition that he would not be sold.

During the first night with the new people (his third) he was accompanied by the owner who slept on the ground near him in a sleeping bag. The owners exercised him regularly and took him to a veterinarian for treatment of an ear infection. At the vet's office, he weighed in at 140 lbs.

The children petted and played with him and he seemed to like two friendly cats as well as two dogs chained to kennels behind the house. By comparison, his life with the previous owners must have been a lonely one. I suggested that perhaps he would once again sleep on the bed if allowed in their trailer home. The owner told me they had already tried this but instead of climbing up on the bed with all his 140

The Management Imperative and Mythology of Animal Rights

lbs, he lifted his leg and urinated on the bedpost which did not meet with approval.

For several months I visited Rio regularly from our cottage in Qualicum Bay. When I pulled up in the vehicle he would bark at first and then slowly wag his big heavy tail. I think he came to recognize me as the person with a handful of dog biscuits in his pocket. He approached me slowly and cautiously from his kennel rather than excitedly as dogs do when offered "cookies." He took these from my fingers gently enough after picking up one or two I had cautiously placed on the ground.

I pressed the owners to try and get more details from the previous Canadian owner, but without success. This leads me to believe that Rio's removal from a wilderness den, and the details of his years in Oregon and those in BC were not for public knowledge. I later learned that it was illegal to possess a wolf in the United States which may have been the reason for the new home in B.C.

Rio (age 8 years), August 2002 (collar and chain not showing).

Unlike domestic dogs, Rio howled periodically and emitted a different skin odour. He was not neutered but apparently not aggressive to other dogs. On one occasion, however, when he was off the leash up the Big Qualicum River a golden retriever dashed away when it saw him. This apparently triggered Rio to give chase and grab the fleeing dog by the neck and force it to the ground. Rio's owner quickly caught up and made him release his hold on the dog's neck. The retriever instantly disappeared into the forest and ignored both owners efforts to call it back. Reportedly, it was 2 hours before it returned to the road. Apparently, there were no visible injuries or behavioural aftereffects.

I believe Rio sensed that I was not fully relaxed when I fed him biscuits. Even when the owner was present I tried several times to pat him on the head but he would back away as though he didn't trust me. For people who approached in the usual casual way we do with domestic dogs Rio reacted in a friendly manner letting them touch or pet him. It seems that he sensed my caution and behaved in a less forward manner. Rio's owner was quite surprised that he reacted differently towards me.

After several months of visits, on the last occasion, Rio was not there. I hastily contacted a neighbour to find that he had been taken away by the SPCA. He had broken loose and allegedly caused an elderly woman to fall and scar her face. The woman's daughter phoned the SPCA and reported that her mother had been brutally attacked.

Next day I phoned the SPCA and was informed that Rio had been adopted out. I asked where he was and informed that such information is not given out. I was invited to visit the office and tell them about my interest in Rio. I was there the next day and learned the true story about the "attack" on the elderly woman. The daughter had not reported the incident correctly. There was no attack. The woman was walking her small dog on a retractable leash when Rio rushed up like most male un-neutered dogs do when seeing another dog. The little dog was frightened and ran to its owner. It wound the leash around the woman's legs, causing her to fall over and scar her face. When she later learned that her daughter had described the incident as a brutal attack, she informed the SPCA of the truth and said she did not want

The Management Imperative and Mythology of Animal Rights

Rio blamed for the incident. This probably saved his life.

At the SPCA, Rio was neutered, had his dew claws removed, and further examined by a veterinarian. He was separated by only a wire fence from the large holding area where the dogs are kept. He showed no aggression towards the dogs. He was later admitted to the general holding area along with the dogs and again showed absolutely no aggressive tendencies. I was told he got along well with all of them.

Rio had been adopted out to a couple with a small boy and a part shepherd female. They would not tell me where he was but offered to phone the owners and tell them about my background and interest in wolves. Subsequently, I was given the phone number and address of the new owners, the fourth in Rio's life.

Rio was living a happy life of freedom with his female part Shepard companion Galena on 15 acres of privately owned forested farm land. The owners were responsible people with an intelligent interest in his welfare. They later recorded on tape Rio's wolf howl. At 4 weeks of age, when removed from his mother, he had probably never heard his parents howl. Apart from his age, this may have accounted for the somewhat high pitch to his howl. Galena also learned to howl.

When I first visited Rio's new home, he and Galena began barking and running toward the vehicle. For the first time, I was able to pat him on the head. It was difficult, however, to feed biscuits to Rio unless Galena was tied up. Unlike Rio the wolf, Galena was anxious to be petted and made a fuss over. When finished eating, Rio was content to wander off and lie down, leaving Galena to seek attention. I realized that Rio was now over 10 years of age. This may also have been the reason he squatted like a female when there was nothing close to urinate against.

My last visit to see Rio was in August 2002. A month later, in September he suddenly became ill and died. It was a sad day for all of us who knew this remarkable animal. It is unfortunate that the first 5 years of Rio's life is not documented. I suspect that his treatment as a house pet during this time was responsible, at least in part, for his behaviour in later years. Ron and Cathy Oliver, who owned him for the last two years of his life, gave him the attention and love he apparently had as a puppy.

The author with Rio – age 11 years (no collar or chain).

Eight-year old Emery Oliver with Rio (11 years) in August 2002, a month before he died.

This was a success story in which a wolf was kept as a pet for at least 11 years. At age 2 or before, they may have to be confined to captivity or even destroyed if aggressive. In the wild, Rio would have been lucky to have lived 6 to 8 years.

Researchers believe there are more than 100,000 captive wolves and 400,000 hybrids in the United States alone (Hope 1994). Others estimate the number of privately owned wolves or hybrids at 8,000 to 2 million (Kramek 1992).

The US Endangered Species Act of 1993 forbids ownership of pure wolves but hybrids are subject to little, if any, regulation. Pet wolves and wolf-dog hybrids are said to have killed at least 9 children in the US from 1986 to 1994 and many others were maimed (Hope 1994). Mivert (1890; cited in Mech 1970) reported that 161 people were killed by wolves in Russia in 1875 alone.

ISBN 1-41206147-4